Teaching Electromagnetics

Teaching Electromagnetics

Innovative Approaches and Pedagogical Strategies

Edited by
Krishnasamy T. Selvan
Karl F. Warnick

CRC Press
Taylor & Francis Group
Boca Raton London New York

CRC Press is an imprint of the
Taylor & Francis Group, an **informa** business

First edition published 2021
by CRC Press

6000 Broken Sound Parkway NW, Suite 300, Boca Raton, FL 33487-2742
and by CRC Press
2 Park Square, Milton Park, Abingdon, Oxon, OX14 4RN

CRC Press is an imprint of Taylor & Francis Group, LLC

ISBN: 978-0-367-71088-0 (hbk)
ISBN: 978-0-367-71057-6 (pbk)
ISBN: 978-1-003-14923-1 (ebk)

Typeset in Times
by SPi Global, India

To my wonderful wife Sudha in deep admiration

Krishnasamy T. Selvan

To Shauna, my loving and supportive eternal companion

Karl F. Warnick

Contents

Acknowledgments

This book in your hands or on your screen presents the experiences and perspectives of a multi-institutional, multinational authorship on the topic of electromagnetic education. The editors are grateful to the authors of the chapters for their great support and participation in this first-of-its-kind project.

Special thanks go to Hugo Espinosa and Cynthia Furse for enthusiastic help and suggestions whenever the editors sought them. As the project encountered obstacles and challenges, their support was invaluable.

We recognize the IEEE Antennas and Propagation Society for hosting the supplementary material for the book and for providing a vibrant community where electromagnetics education is a welcome part of the discussion.

Marc Gutierrez and Nick Mould of the CRC Press were always ready with appropriate clarification as we dealt with the many matters consequent to a project like this. Our sincere thanks are due to them.

Finally, no major undertaking in life, including this book project, would be possible without the loving support of one's family. Krishnasamy Selvan thanks his wife Sudha, children Srinidhi and Shyam Krishna, and mother Dhanalakshmi Krishnasamy for their admirable support and understanding. Karl Warnick thanks his wife Shauna and their six children for their support and encouragement.

Contributors

Hugo G. Espinosa
School of Engineering and Built
Environment
Griffith University
Brisbane, Queensland, Australia

Thomas Fickenscher
Faculty of Electrical Engineering
Helmut Schmidt University
Hamburg, Germany

Cynthia Furse
Department of Electrical and Computer
Engineering
University of Utah
Salt Lake City, Utah, USA

Jenny James
Educational Designer
Learning Futures
Griffith University
Brisbane, Queensland, Australia

Uday Khankhoje
Department of Electrical Engineering
Indian Institute of Technology Madras
Chennai, Tamil Nadu, India

Nickolas Littman
School of Engineering and Built
Environment
Griffith University
Brisbane, Queensland, Australia

Soo Yong Lim
Department of Electrical and Electronic
Engineering
University of Nottingham
Selangor, Malaysia

Anthony A. Maciejewski
Department of Electrical and Computer
Engineering
Colorado State University
Fort Collins, Colorado, USA

Sanja B. Manić
Department of Electrical and Computer
Engineering
Colorado State University
Fort Collins, Colorado, USA

Ryan McCullough
Department of Electrical and Computer
Engineering
Colorado State University
Fort Collins, Colorado, USA

Branislav M. Notaroš
Department of Electrical and Computer
Engineering
Colorado State University
Fort Collins, Colorado, USA

Berardi Sensale Rodriguez
Department of Electrical and Computer
Engineering
University of Utah
Salt Lake City, Utah, USA

Yana Salchak
School of Engineering and Built
Environment
Griffith University
Brisbane, Queensland, Australia

Krishnasamy T. Selvan
Department of Electronics and
Communication Engineering
Sri Sivasubramaniya Nadar College of
Engineering
Kalavakkam, Tamil Nadu, India

Levent Sevgi
Istanbul OKAN University
Turkey

Eng Leong Tan
Nanyang Technological University
School of Electrical and Electronic
  Engineering
Singapore, Singapore

Javier Bará Temes (retired)
Polytechnic University of Catalonia
Barcelona, Spain

David V. Thiel
School of Engineering and Built
  Environment
Griffith University
Brisbane, Queensland, Australia

Karl F. Warnick
Department of Electrical and Computer
  Engineering
Brigham Young University
Provo, Utah, USA

Donna Ziegenfuss
Marriott Library
University of Utah
Salt Lake City, Utah, USA

1 Introduction

Krishnasamy T. Selvan

Sri Sivasubramaniya Nadar College of Engineering, India

Karl F. Warnick

Brigham Young University, USA

CONTENTS

1.1 PREAMBLE

This book is concerned with effective teaching and learning of electromagnetics. Electromagnetic (EM) theory is part of the foundation of electrical engineering, wireless communications, signals and systems, optics, nanotechnology, and many other modern technological application areas. It has broad significance, since mastering electromagnetic fields and waves can help with acquiring "quite strong premises to understand also other fields of engineering, for example hydraulics" [1]. It is naturally a core subject in most universities for undergraduate electrical engineering and physics programs.

Not withstanding the importance, as the state-of-the-art moves forward in engineering and applied sciences, traditional academic topics like electromagnetics are squeezed into smaller curricular spaces to make room for design projects, writing and communication skills, and up-and-coming engineering application trends. Student expectations for the college experience and methods for learning and curriculum delivery are changing. After many decades of evolution in the teaching methodologies, student perspectives, and curricular structure of the field theory content in undergraduate education programs around the world, now may be a good time to pause and consider where teaching is at in this field and where it should be headed in the next few years. How should teachers and learners of electromagnetics respond to these pressures and opportunities?

Given the challenges and complexity of the teaching-learning process, there is no one-size-fits-all approach for all subjects. Complexity arises because of asymmetry in the nature of the various subjects and in student learning styles. According to the

1

noted educational theorist David Kolb, effective learning occurs through a four-stage cycle that involves concrete experience, reflective observation, abstract conceptualization, and active experimentation. While each of these modes helps in the teaching and learning process, considering that engineering has traditionally been taught by introducing a concept or theory (conceptualization) and then following it up with discussion of problems (experimentation), learning can begin with any of the four steps [2]. Since classrooms are places where learning in a sense *begins* and gets nurtured, it can be hoped that any of the four steps, or a combination of them, when effectively employed, can lead to a well-rounded teaching and learning experience.

Modern educational challenges and developments warrant that educators attempt varied methods for effective teaching and learning. Student attention spans and learning styles are evolving, as is our understanding of the science of teaching and learning [3]. Active learning, the flipped classroom, and remote teaching during a period of public health challenges all raise questions for the instructor as to how to incorporate new methods and teaching styles in the classroom.

This book is envisioned to be a guide and reference book for electromagnetic educators, addressing the above questions and challenges about curricular content and pedagogical methods primarily at the undergraduate level. Topics include teaching methods and lab experiences and hands-on learning. The book grapples with issues related to online instruction, which is timely given the recent worldwide shift to remote teaching. While many instructors will likely return to in-person instruction, some of the necessary adaptations will become permanent. We hope to give the educators ample food for thought with respect to evolving teaching methods based on the experiences of those who have expertise in these areas specific to electromagnetics teaching. The end result should be impact on the reader represented by improvements to his or her practical teaching methods and curricular approach to electromagnetics education.

1.2 EDUCATIONAL APPROACHES IN FOCUS

Much of this book is built around two essential educational aspects in respect of effective teaching and learning of electromagnetics: experiential learning and conceptual learning, two of the modes in Kolb's cycle that can also encompass a third mode, reflective observation.

Experiential learning helps with the promotion of visualization, experimentation, and interpretation skills and in turn facilitates "an effective construction of knowledge" [4]. While real experimental set-ups have their own charm, when financial or other constraints impose restrictions, virtual experiential learning can be employed. Virtual experiential learning is "achieved through the integration of different motivating technologies like virtual reality, augmented reality, games and simulations in an active learning context" [4].

Conceptual learning acquires significance in the context of one of the main dilemmas facing engineering education, striking "the balance between immediate and long-term skills" [1]. While engineering students ought to learn various important skills including design tools and standards, since their "validity may expire over the technological development," gaining "profound understanding and readiness for the changes in rapidly evolving industry" requires the development of long-term engineering skills and a strong foundation in the core subjects [1]. For electrical engineering students, electromagnetic theory being an important core subject, instilling a strong conceptual understanding in the subject is a priority for educators. Students' ability to develop content knowledge in subject can be significantly impacted by the teacher's interest and enthusiasm in delivering it [5].

Besides focusing on experiential and conceptual learning aspects, the book presents ideas on expanding the scope of electromagnetic education. Examples include teaching the subject to computer science students and medical technicians, and on implementing the flipped classroom approach for EM. The opportunities and challenges presented by disruptive times to education are considered. We hope to combine timely aspects of the current learning environment with the timelessness of the theoretical fundamentals in a stimulating discussion on teaching and learning electromagnetics.

1.3 ORGANIZATION

Within the above framework, this book is organized into thirteen chapters grouped under the following topical areas:

- Introduction and philosophy (Chapters 1 and 2)
- Experiential learning – real and virtual (Chapters 3 to 6)
- Conceptual learning (Chapters 7 to 9)
- Teaching approaches (Chapters 10 to 12)
- Conclusion and outlook (Chapter 13)

Chapter 2 undertakes a high-level philosophical consideration of the subject, as it would be an appropriate starting point. This chapter considers the issues, challenges, and pressures that face those who teach electromagnetic theory. Taking into account changing student learning styles and the desirability of expanding the scope of the subject beyond applications, it identifies some of the trends and opportunities that lie ahead, and make recommendations for the next decade of training and course development in the field of electromagnetics.

In Chapter 3, an experiential learning approach to electromagnetics courses is presented. The structure of the course along with examples of student work are presented, and the effectiveness of the learning approach as evaluated by means of student feedback discussed.

Chapter 4, which is reprint of [6], considers the use of MATLAB for effective teaching of electromagnetics. Providing an introduction to students such that they can write their own codes for electromagnetic fields, the chapter demonstrates how the approach can powerfully impact learning and enquiry. The chapter also discusses results of student feedback obtained on the effectiveness of the method.

Chapter 5 describes a range of interactive computational MATLAB-based tools for assisting in teaching electromagnetics. Such tools aim to improve student engagement, learning outcomes, reduce staff workload, and increase the student interest in electromagnetics education.

Chapter 6 presents computational methods and mobile apps for electromagnetics education. These mobile apps feature convenient and effective touch-based interactivity to aid teaching and learning. Some educational study and survey results are also presented.

Chapter 7 discusses the advantages of using the calculus of differential forms in teaching EM theory. Developing EM theory and the calculus of differential forms in parallel from an elementary, conceptually oriented point of view using simple examples and intuitive motivations, the chapter concludes that the power of the calculus of differential forms provides an attractive and viable alternative to the use of vector analysis in teaching electromagnetic field theory.

Chapter 8 suggests that the concept of Maxwell's displacement current, if taught in the context of its development, can help with inculcating the qualities of innovation and creativity in students. To this end, the chapter presents a brief account of the development of the concept and discusses two fundamental questions addressed in the literature. Suggesting a teaching approach, the chapter reflects on how such a delivery can be consistent with the broader goals of higher education.

Chapter 9, a reprint of [7], describes the rationale of a one-semester, sophomore level, basic course on electromagnetic waves for electrical engineers, with emphasis on communications. The approach avoids the mathematical complexity associated with Maxwell's equations but without any significant loss of rigor.

Chapter 10 presents the authors' efforts to take electromagnetics to non-EE domains. Programmes for computer science students and medical technicians are considered in this regard.

Chapter 11 discusses Hybrid-Flexible (HyFlex) course models where students can take a course in person and/or online, and seamlessly move between the two modalities. The chapter describes ways to create a student-centered HyFlex course using a Flipped teaching model (HyFlexFlip or HFF).

Chapter 12 brings together the experiences and reflections of academics in differing geographical regions, in the context of the sudden shift to virtual learning in response to COVID-19.

Chapter 13 reflects on the outlook for electromagnetic education and concludes the book.

REFERENCES

1. A. Korpela, T. Tarhasaari, L. Kettunen, R. Mikkonen and H. Kinnari-Korpela, "Towards deeper comprehension in higher engineering education: method of cornerstones," *European Journal of Science and Mathematics Education*, vol. 4, no. 4, pp. 418–430, 2016

2. D.A. Wyrick and L. Hilsen, "*Using Kolb's cycle to round out learning*," *Proceedings of the 2002 American Society for Engineering Education Annual Conference & Exposition, Session Number 2739.* Available at: https://peer.asee.org/using-kolb-s-cycle-to-round-out-learning, accessed February 10, 2021

3. C. Brown, H.L. Roediger III, and M.A. McDaniel, *Make It Stick: The Science of Successful Learning*, Cambridge, Massachusetts: Harvard University Press, 2014.

4. C. Vaz de Carvalho, *"Virtual experiential learning in engineering education,"* *2019 IEEE Frontiers in Education Conference (FIE)*, Covington, KY, USA, 2019, pp. 1–8, doi: 10.1109/FIE43999.2019.9028539.
5. M.M. Keller, K. Neumann, and H.E. Fischer, "The impact of physics teachers' pedagogical content knowledge and motivation on students' achievement and interest" *Journal of Research in Science Teaching*, vol. 54, no. 5, pp. 586–614, 2017.
6. B.M. Notaroš, R. McCullough, S.B. Manić, and A.A. Maciejewski. Computer-assisted learning of electromagnetics through MATLAB programming of electromagnetic fields in the creativity thread of an integrated approach to electrical engineering education. *Computer Applications in Engineering Education*, vol. 27, pp. 271–287, 2019. doi: 10.1002/cae.22073
7. J.B. Temes. Teaching electromagnetic waves to electrical engineering students: an abridged approach. *IEEE Transactions on Education*, vol. 46, no. 2, pp. 283–288, May 2003, doi: 10.1109/TE.2002.808275.

2 Teaching and Learning Electromagnetics in 2020

Karl F. Warnick

Brigham Young University, USA

Krishnasamy T. Selvan

Sri Sivasubramaniya Nadar College of Engineering, India

CONTENTS

2.1 INTRODUCTION

This chapter is based on an article by Warnick and Selvan of the same title in the *IEEE Antennas and Propagation* Magazine's 2020 special issue on education.

Electromagnetic (EM) theory is a quintessential academic discipline. It includes abstract and challenging mathematical concepts such as vector fields and the curl operator that have required concentrated effort from students of electromagnetism for the last 150 years to learn and comprehend their deeper meanings. It is a coherent body of material with a pleasing structure and a natural pedagogical progression from circuits and statics to fields and waves. It connects mathematics and physics and is used in a wide variety of engineering applications, including digital systems, bioengineering, power and energy systems, optics and optoelectronics, high-frequency devices and circuits, microwave systems, remote sensing, wireless communications, and medical imaging.

The significance of EM today has been well articulated by many prominent experts in the field:

Stephen A. Boppart, Departments of ECE, Bioengineering and Medicine, UIUC [1]: "Light, and its interactions with biological tissues and cells, has the potential to provide helpful diagnostic information about structure and function. The study of EM is essential to understanding the properties of light, its propagation through tissue, scattering and absorption effects, and changes in the state of polarization."

Andreas C. Cangellaris, ECE Department, UIUC [1]: "One of the most intriguing, rewarding and challenging experiences of my academic career is the teaching of the fundamentals of EM fields and waves to undergraduate electrical and computer engineering (ECE) students. What makes it intriguing is the fact that it is these concepts that every ECE student will rely upon as he [or she] tries to think through and comprehend the basic principles behind the operation of each and every electronic device, component, circuit or system that constitute the building blocks or the enabling force of the electrical power, communication and computing revolutions of the past century."

Fawwaz Ulaby and Umberto Ravaioli: "We are constantly surrounded and bombarded by electromagnetic waves, some visible—including sunlight, starlight, and every type of indoor and outdoor light we use every day—and many others that are invisible to our eyes but not to the antennas in our cellphones and WiFi routers everywhere" [2].

While the pedagogical beauty of electromagnetic theory and applications such as these provide a compelling reason to learn electromagnetics, those who teach in this discipline face challenges. Educational trends, technological advances, and new directions and developments can make courses in classical field theory seem old-fashioned. Faculty in adjacent subdisciplines see the credit hour allocations to fields and waves as easy targets for trimming. Field theory courses are reduced to make way for solid-state physics, embedded programming, optics or microwave engineering courses, or topics like writing, ethics, and leadership.

When faced with pressure to reduce coverage, our reflex is commonly to fight to protect and preserve cherished course material. This is in line with the traditional role of the scholar. Researchers have the goal of creating new knowledge, to be sure, but the scholar also seeks to preserve the wisdom of the past, keep the flame of deeper truth alive, and to light the spark of appreciation for higher knowledge in the hearts and minds of students. It can feel like the ultimate insult when one's disciplinary field is minimized in importance and pushed from the curriculum by administrators and faculty in other areas.

As our knowledge expands, we must evolve the curriculum. Teachers of electromagnetic theory must weigh the desire to be good stewards of the discipline against the need to evolve and improve the curriculum. Learning Latin years ago was a mind-expanding rite of academic passage, but most would agree that we are well served by using time once devoted to that topic for other purposes within the typical university general education.

Faculty consequently feel a natural pressure to replace course content and improve efficiency in the curriculum. Even as science and technology grow more complex, educators hope students can be brought reasonably close to the state of the art at the end of a four-year college degree. This means that faculty must adjust and prune course material in all foundational topics, including electromagnetic theory. Student learning styles and habits change, especially as technologies and delivery mechanics used in education evolve, and this should be reflected in the way classical topics like electromagnetic theory are taught.

With these guiding principles in mind, we will make recommendations for teaching and learning in electromagnetic theory over the next decade in these areas:

1. How to focus the topical coverage to fit with a well-crafted curriculum in electrical engineering or similarly constituted degree programs;
2. How to facilitate experiential learning by incorporating numerical methods and visualizations;
3. How to address changes in modes for course material delivery and student learning styles; and
4. Long-term objectives and the different regional needs of students as they are relevant to electromagnetics teaching.

The goal is to provide a framework that teachers and curriculum developers can use in thinking about how to respond in a constructive way to pressures and changes associated with electromagnetics teaching and learning in the next decade. Our focus is the electromagnetics-related curriculum in a college or university undergraduate degree program in electrical engineering, electronics engineering, communications engineering, and similarly constituted degree programs.

2.2 TOPICS AND COVERAGE

One of the long-term needs in education is to find ways to present earlier material more simply and efficiently and to make room for new topics. This inevitably happens with electromagnetics education. A wave of course reductions has already occurred over the last generation. Most university programs use a leaner approach than a generation ago when teaching field theory.

A typical undergraduate electrical engineering program includes preparatory math courses that include multivariable calculus and vector analysis, a physics course that touches on static fields, and a course on electromagnetic fields and waves, usually including transmission lines. Near the end of a four-year degree, students select from elective courses on optical systems, antennas and propagation, or microwave engineering and high-frequency circuits. The curriculum may include a class that emphasizes antenna analysis and wireless communications or the coverage may focus on materials, guided waves, and modes. Some programs emphasize optics, devices, and physical electronics, or integrate the treatment of waves with quantum effects.

This kind of variety in emphasis is unavoidable, given the diversity in preferences and viewpoints on the subject and its importance in our community. The level of students coming into a program too places significant constraints on what can be profitably be covered and what cannot. There is perhaps no possibility for a 'one-size-fits-all' approach to framing curriculum.

We often encounter practicing RF engineers who feel that they have never "solved Maxwell's equations" in the course of their work. They use software tools to design transmission line structures, high-frequency circuits, and antennas and they question the value of learning electromagnetics from a rigorous mathematical perspective.

These issues cannot be lightly passed over. It is easy for older, experienced teachers of EM theory to simply restate the platitude that all engineering students should learn Maxwell's equations and have a rigorous grounding in vector calculus. Yet one can easily envision a full curriculum with only a light touch on Ampere's and Faraday's laws and a quick transition to microwave engineering, transmission lines, and a practically oriented perspective on antennas as just one of many components in a communications system.

A purist would argue that no student has truly learned electromagnetic theory without understanding modes. A practitioner might well counter that any number of powerful circuits, systems, and structures can be developed and designed using only the transverse electric and magnetic (TEM) mode and the basics of transmission line theory.

What is the "right" approach – and is there a "right" approach to teaching university-level electromagnetics? What are the most important aspects of EM theory that should not be missed in an effective treatment of the subject? One could argue that there are critical topics that all electromagnetics curricula must cover in order to be complete and credible, but the subset of critical topics tends to expand rather than converge as the number of experts providing input in the discussion increases. In a modern accredited program, the topics covered might be dictated by the regional needs of students and the interests of program stakeholders. These considerations illustrate the challenge of structuring an electromagnetics curriculum within a framework of limited class time and motivate the need for ongoing reassessment of the topical coverage and course sequence.

2.2.1 UNDERGRADUATE CURRICULAR MODELS

Some undergraduate electrical, electronics, or communications engineering programs with a strong tradition in applications of EM theory allot two full courses to classical fields and waves before moving to elective content. In a two-course sequence, the first course might cover statics, power and energy, Maxwell's equations for time-varying and time-harmonic fields, plane waves, propagation in dielectrics and conducting media, and transmission lines. The second course could include plane wave reflection and transmission, radiation, antennas, Friis transmission formula and link budgets, simple propagation models, resonant cavities, and waveguides.

The two-course sequence allows deeper development of EM theory and a broader coverage of traditional topics than a single EM course. In many programs the pressures discussed above have led to reduced coverage and hard choices in the selection of topics. The two-course sequence is often chosen by programs that prepare a large proportion of students for graduate degrees in electromagnetics and its applications. University faculty often find it difficult to balance the competing demands of preparing BS level students for industry positions and the greater rigor and mathematical depth needed for prospective MS and PhD students.

Curricular limitations at the undergraduate level can be mitigated in first-year graduate courses, particularly when the graduate research emphasis is on modern, applied topics such as optical systems or array antennas rather than classical field theory. Many instructors find that graduate students require remedial coverage of undergraduate topics in EM theory.

Many programs have one course on electromagnetic theory followed by elective upper division or higher level courses on applied electromagnetics topics. A typical flow for a one-semester undergraduate course on electromagnetic theory is as follows: Overview of applications to provide motivation – Broad picture of EM theory – static fields – electromagnetic fields – propagation in lossless and lossy media – reflection and transmission – radiation and antennas.

The one-course structure assumes that the students have had a mathematics course that includes calculus and vector analysis in an earlier semester and a touch on electrostatics or magnetostatics in a physics course. If the EM course is completed in the second or third year, antenna and wave propagation and microwave circuit design can be offered as elective courses in higher semesters. This is the recommendation in a recently suggested model curriculum in India for undergraduate program in electronics and communication engineering [3].

In a practically focused, "maker"-oriented curriculum, the priorities might be a short coverage of quasistatics to develop the basics of inductance and capacitance and complement circuits courses, followed by transmission line theory, impedance matching, S-parameters, radiation from simple antennas, and link budgets. This approach would include less coverage of plane waves, reflection and transmission, and other classical solutions to Maxwell's equations than the models discussed above, but better prepares students to use modular RF communication systems in project-based learning classes or take a follow-on circuits and systems focused microwave engineering course.

With any choice of topics, there are challenges. The one-course model does not have a natural place for more than a brief touch on waveguides and modal theory. This and other more advanced topics in boundary value problems displace practically focused material on circuits or a deeper coverage of antenna theory. Guided wave structures have taken a back seat to microstrip and printed circuits in most application areas, so waveguides and modal theory might be less important today than a solid coverage of transmission line theory and antennas.

2.2.2 Broader Considerations for EM Curriculum Content

These considerations highlight the typical conundrum facing educators in EM theory of fitting topics into a one- or two-course sequence in a way that respects traditional theoretical and mathematical treatments while preparing students for relevant industrial applications. Any specific curriculum represents a crystallization of views on the relative importance of topics from EM theory. If the EM theory instructor has thought through the issues, is well prepared to engage in a conversation with other faculty, and understands the need for give and take as the topical mix evolves over time, chances are that good decisions will be made about the curriculum.

In their own programs, the authors have seen a gradual shift over the past two decades from theoretical depth to practical applications of electromagnetics with less coverage of classical topics. Much of the funded research in the U.S. and other countries have moved away from theory and mathematical topics toward systems and applications, so the shift in teaching focus parallels changing research priorities.

An area for more efficient treatment is to better integrate undergraduate engineering electromagnetics with preparatory mathematics and physics courses. Based on tradition and the need to manage resources and faculty in reasonably sized groups, universities have generally retained departmental boundaries between mathematics, physics, and engineering disciplines. Greater innovation in the creativity and effectiveness of electromagnetics training could be fostered by increased collaboration between engineering, mathematics, and physics faculty.

2.3   NUMERICAL METHODS AND VISUALIZATIONS

We have discussed the tension between deep coverage of fundamentals and rapidly moving to applications in the undergraduate EM curriculum. There is also tension between mathematical and theoretical foundations and numerical simulations and visualizations, motivated by the argument that the latter helps with experiential learning of the subject

We might consider two extremes in schools of thought on numerical simulations and visualizations in EM education. One perspective rejects the wide use of numerical models, arguing that students must deeply understand the mathematical foundations in order to become competent in the discipline. Only when students have achieved facility with the theoretical tools can they reasonably be expected to use numerical methods properly and interpret results correctly.

Another extreme would be to view overly exhaustive treatments of theoretical details as archaic and unneeded. Armed with modern software tools, engineers can design microwave circuits, antennas, and full systems with no knowledge of vector calculus. From this viewpoint, students are well served by seeing finite difference time domain simulations of wave propagation, radiation, and diffraction. The visual intuition provided by this approach may stay with students longer than classroom-focused analytical techniques and better prepare them to grasp the complexities of wave propagation as they deal with real-world design situations.

How do we resolve this conflict between mathematical and numerical approaches in the next decade? One approach is to formulate the curriculum in accordance based on the instructor's background and perspective. Cheng, in his widely used textbook [4], states that "I have given considerable thought to the advisability of including computer programs for the solution of some problems, but have finally decided against it. Diverting students' attention and effort to numerical method and computer software would distract them from concentrating on learning the fundamentals of electromagnetism."

Noting that modeling and simulation has basic theory as its conceptual basis, [5] recommends a "balance to be maintained between teaching essentials (theory) and cranking the gear (blind computer applications)." Matthew Sadiku has noted that "vector analysis is the language of electromagnetics which needs to be mastered well" [6]. It is hard to imagine entirely abandoning one of the most useful and successful mathematical frameworks ever created – vector analysis – entirely in favor of numerical visualizations.

Fawwaz Ulaby and Umberto Ravaioli explain it this way: "when teaching electromagnetics to students, it is important to strike a balance between the rigorous mathematics required to understand how EM waves propagate through and interact with different media, and the many direct applications of electromagnetics: communication systems, radar, medical applications, RFIDs, GPS, and liquid crystal displays, among many others" [2].

One approach to resolving this tension is to have students implement numerical methods. Running a standard set of computer programs and looking at the field variations in a printed antenna can to some extent help with appreciating what goes on in the antenna, but may divert the focus from understanding deeper concepts behind its operation. The student will often have little idea as to how those fields could be predicted in the first place. When the student writes a simple finite difference code starting from Maxwell's equations, the situation improves. While deriving the technique-specific equations will require an understanding of the underlying theory, writing the code and running it further enhances the understanding of the problem. The structure of a code implementation of Maxwell's equations can illustrate the physical meaning of the equations. Accessible treatments of numerical methods for engineers are available [7].

A balanced approach is to include in the curriculum the development of CEM codes for simple problems embedded in the mathematical development. Examples include analysis of free space propagation with a dipole radiator or 1D pulse propagation and reflection. When students develop their own code and use them to visualize the radiated fields, they are better prepared to make more effective use of commercial codes for visualizing the behavior of other structures and also for designing them. More advanced visualizations can be provided by the instructor. In summary, we feel that visualizations and numerical code development support but do not replace a mathematical treatment of field theory.

2.4 CHANGING LEARNING STYLES

As the role of technology in daily life has expanded and evolved, students have changed in their learning styles and habits. Some prefer to watch videos of lectures rather than attend class. Content delivery takes place across a diverse set of platforms, from the classroom to mobile devices. Students expect more help and close interaction with teachers. They are eager to learn but are frustrated by topics that they cannot quickly understand. Students are motivated by grand challenge problems at the societal level and want to design and make things but can lose interest when course material becomes tedious. How do we adjust electromagnetics teaching to reflect these trends?

Personal qualities of the teacher – enthusiasm, love of teaching, concern for students, and commitment to the discipline – will always be valuable. The teacher can help education transcend the raw, antiseptic transmission of information. These intangible factors give traditional instructor-led courses an effectiveness that is hard to replicate with pre-recorded videos, massive online courses, or other less personal delivery mechanisms. While the latter modes of delivery are probably suitable for continuing education, for regular mainstream education the value added by a competent and enthusiastic teacher is not replaceable.

An enthusiastic teacher can elevate the level of student interest even in what are perceived to be subjects that are either difficult or practically irrelevant or both. As one of the authors has said, "An effective teacher must be passionate about their subject. Only if you are enthusiastic will the students have the chance to develop enthusiasm themselves" [8]. When the teacher is positive about the teaching and learning process the attitude can be contagious and can significantly help with student learning.

Teachers must adjust to the widely acknowledged observation that student attention span is declining. Interest can be retained by inserting activities in the lecture. These can include experimental demonstrations, discussing interesting and related historical anecdotes, asking students to make presentations on related topics of their choice, short quizzes, and evaluations, or brief writing assignments.

Students can be energized by a teacher who delivers lectures through the semester on important topics in historical context. The topic of displacement current has been considered in this context [9]. Surveys of students suggest that context-based lecturing enhances interest [10–12].

Many university programs are beginning to shift attention from delivering content to improving the skills and knowledge that are actually acquired by students or outcomes-based education. This moves the focus from teaching to learning. Pressure to improve student outcomes has stimulated a healthy and valuable conversation among educators. In the U.S., outcomes-based educational theory became prominent in engineering when the major accrediting body shifted the context of evaluating engineering programs to ABET EC2000. The earlier approach to accreditation (evaluation of facilities, faculty, and credit counting of topical coverage) was augmented with a framework for assessment, evaluation of student learning outcomes, and continuous improvement. Most programs responded to the change reasonably well. Complaints about the additional burden of the new accreditation regime [13] and subjectivity among external program evaluators in reviewing assessment strategies in accredited programs have led in the years since 2000 to adjustments of the extent of outcomes and assessment based program accreditation.

A parallel revolution has occurred with the proliferation of online, video-based learning platforms like Udacity, Coursera, the Khan academy, and many others. After initial enthusiasm, it was found that online programs have low completion rates. Despite far lower cost structures, they have not revolutionized higher education as rapidly as its proponents have predicted. On the positive side, students routinely turn to online resources as a supplement to traditional bricks and mortar educational programs. Some fully online programs have survived and even flourished.

Faculty teaching electromagnetics should be aware of the expanding body of online resources for EM theory and supporting mathematical topics. The authors would welcome more data on how widely used or effective online resources are for students in EM courses and invite instructors with experience in incorporating online resources in EM courses to share their experiences with the community. Some of these experiences are found in other chapters in this book.

The persistence of traditional educational models even in the face of rapid technological evolution suggests that it is hard, even with technology advances, to improve on what has crystallized over centuries of experience with university education. The current mix of lectures, assignments, and exams used in teaching EM and many other topics might be criticized as outdated, but it does work well for many students and represents a reasonable balance of effectiveness with the time available for the typical academic to divide between teaching and other responsibilities. It is surprisingly difficult to find something that is uniformly better in all respects than traditional lecture-based content delivery.

Despite challenges, improvements are always possible and the quest for growth should be unceasing. The excellent summary of the science of teaching and learning in the book *Make It Stick* [14] has stimulated many faculty to improve teaching methods in a research-based way that focuses on long-term retention. Requiring students to recall material from several weeks earlier (retrieval practice), interleaving topics, embracing the difficulty of learning, and other learning methods have had an impact at the authors' institutions.

A related movement has been the "flipped classroom." Students learn the material from videos or other materials before class then work problems in the classroom with assistance from the instructor. Cynthia Furse at the University of Utah and Mike Potter at the University of Calgary, among others, are members of the IEEE Antennas and Propagation Society who have applied these techniques extensively in engineering and electromagnetics education [15]. EM theory, being a challenging topic with abstract principles that are perennially hard for students to grasp, promises to be a fertile proving ground for active learning strategies like the flipped classroom.

Even with innovation in delivery methods, there is still a need for capable and inspiring individuals to teach and train students in EM theory with a personal touch. Raghunath Shevgaonkar observed that "among the electrical engineering subjects, electromagnetics is the richest in concepts. In most of the other subjects, there are just a few concepts and a wide range of their applications. It is my experience that when EM is taught by teachers who make the subject appear simple and conceptual, the students get inspired and love EM. A subject like EM makes students mathematically strong and imaginative. These skills then help students in creating new innovations even in non-EM subjects" [6].

2.5 EDUCATIONAL OBJECTIVES AND GLOBAL PERSPECTIVES FOR TEACHING ELECTROMAGNETICS

As we have seen, making choices in EM education like many endeavors involves competing principles that seem challenging to balance. Kim B. Clark, former Dean of the Harvard Business School, referred to these as the "trade-offs that seem ironclad." The trade-offs can be addressed, but this requires thought, creativity, and inspired collaboration among faculty and other stakeholders.

One of these trade-offs is the tension between general education and vocational training. We hope that students will leave the university prepared to be employed and to contribute in a valuable way to society. At the same time, education has helped to lift countless numbers of peoples and societies by teaching principles of cooperation, problem-solving, rational thought, mature discourse, and the joy of shared inquiry. Enlightened thinkers help to bring about stable and effective institutions and peace among nations.

According to an Australian Higher Education Council's report, "it is broadly agreed that if higher education is to enable graduates to operate effectively in a range of activities over a period of time, a lifetime in effect and not just immediately after the studies are completed, then it must develop the characteristics that support learning throughout life. Discipline specific skills in many areas have only a short life, and what will be needed in even the medium-term cannot be predicted with any great precision" [16]. In an afterword to the book *The Chicago Handbook for Teachers: A Practical Guide to the College Classroom* [17], the authors write: "One of the rewards of good teaching...should be the knowledge that we have instilled modes of thinking, created intellectual passions, promoted forms of tolerance and understanding, and, of course, increased knowledge."

Considering these aspects, the teaching and learning process, in addition to focusing on the practical and applied dimensions of EM, should also focus on the holistic development of students. While electromagnetics has a practical and applied dimension, at the same time it is an immensely "liberal arts" topic in the sense of its internal logical structure, mathematical beauty, and the historical origins of its key ideas. Therefore, it can be productively used in efforts toward instilling in students holistic qualities.

An example that illustrates teaching with this mindset is Maxwell's displacement current. "Considering the immense scientific and philosophical depth of Maxwell's displacement-current concept, it appears that it would be desirable to make particular use of this concept to enhance teaching and learning. ...Considering that development of science does not often occur in a simplistic way, this would often require that scientific theories be presented along with the context in which they were developed, the methodology adapted, and an appropriate historical account. It has been suggested that such a presentation may entail certain additional benefits, such as acting as a motivator, facilitating intellectual openness, and making the lecture interesting" [9]. A survey taken among students and professors tended to agree with this perception [10,11].

Electromagnetic theory teaching may be employed beneficially in teaching creativity, intellectual spirit, diversity, and life-long learning ability [18]. Electromagnetics is linked to basic principles of physics and is deeply mathematical. This gives room for an instructor who can clearly and enthusiastically teach these difficult topics to inspire students to connect rigorous classical topics to modern applications. It provides a solid foundation for life-long learning and respect for the life of the mind. When the classes are interesting and the instructor enthusiastic students may be able to focus less on extrinsic motivation and more on the joy of learning.

Even when students appreciate the intrinsic value of learning the topic there will always be a need to consider future employability. Because the employment accessible to graduates is driven mainly by regional economies, proper balancing of the trade-off between general education and vocational training is necessarily region-specific.

While an application-based teaching is valuable, it assumes that most, if not all, electrical engineers are employed in core industries. If we consider published statistics and analyses, this assumption is clearly seen to be incorrect. In the United Kingdom, for example, "every year, virtually every grad role is filled – it's just that despite all this talk of skills shortages, the engineering schemes aren't big enough to recruit all the graduates. So they go and do other things because there literally aren't enough jobs for all of them to get in engineering" [19]. Consider this statement from an academic in the UK [19]: "It is astonishing, in the light of claims of science graduate shortages, that so few new graduates go into related employment." According to [20], "about one in five (20%) were employed in roles that were not directly related to their degree and about one in four (24%) were in 'non-graduate' employment, for example working as waiters or in shops."

In India, a large number of electronic engineering graduates come out of colleges every year. Given that the Indian contribution to global electronic industry as of 2011 was only 2.5% (though expected to grow to 15% in 2020) [21], job availability in this sector as at present is inadequate. This means engineering graduates must necessarily find other jobs, such as in the IT sector. In addition to its application-orientation, EM education in this region, as also in others where job prospects in the core sector are not commensurate with the number of graduating engineers, must be considered a component of a modern 'liberal education' as suggested above. As employment opportunities in electronics sector increase, EM education might then follow the same arc as in North America and Europe and become less classical and more application-oriented. Still, the intellectual value of EM theory remains as a benefit to students even as the focus becomes more applied. Further reading on region-specific aspects of EM education can be found in [22,23].

2.6 CONCLUSIONS

We have attempted to lay out in plain view the issues that those who teach electromagnetic theory are dealing with today. The intent is to grapple with conundrums and trade-offs in EM education and to help guide the thoughts of those who must weigh and balance competing pressures on electromagnetics curricula. In the areas considered, we summarize our thoughts as follows:

1. EM curricula are trending toward one theory course that covers fields, waves, and transmission lines, followed by elective courses on microwave engineering or other applications.
2. Toward facilitating experiential learning, numerical methods and visualizations should be brought into the undergraduate curriculum to provide insight into electromagnetic fields without completely displacing the classical mathematical treatment.
3. Bricks and mortar institutions are not likely to be completely replaced by online-only courses for topics like EM theory. Teachers can adapt to changes

in student learning styles by introducing targeted historical background and using principles of learning such as those in [14]. A coordinated approach involving engineering, mathematics, and physics faculty may foster more effective EM teaching and learning.

4. The overall educational experience of students is enhanced by faculty enthusiasm and personal attention. Ways of nurturing these factors should continue to receive our attention.

5. In regions of the world where employment opportunities in core electronic sector are weak, EM teaching can focus on both application-oriented content and technical "liberal education" values that add depth to other more vocationally-focused training and helps build holistic qualities in students.

A shared acknowledgment of these challenges and recommendations might help to produce a sense of solidarity and confidence as we identify and focus on the key aspects of our discipline that should be retained as curricula are modernized and evolve into the next decade in undergraduate teaching programs at institutions around the world. Additional educational resources related to this chapter can be found at the IEEE AP-S Resource Center, https://resourcecenter.ieeeaps.org/.

REFERENCES

1. http://faculty.ece.illinois.edu/rao/EM/Why_Study_EM.pdf, accessed March 29, 2019.
2. F. T. Ulaby and U. Ravaioli, *Fundamentals of Applied Electromagnetics*, Pearson, 2015.
3. https://www.aicte-india.org/sites/default/files/Vol.%20I_UG.pdf, accessed March 28, 2019.
4. D.K. Cheng, *Field and Wave Electromagnetics*, 2nd ed., Reading, Massachusetts: Addison-Wesley, 1989.
5. L. Sevgi, "Modelling and simulation strategies in electromagnetics: Novel virtual tool and an electromagnetic engineering program," available at: http://citeseerx.ist.psu.edu/viewdoc/download;jsessionid=242E13D5F421E310AAE3C42445C2CAEE?doi=10.1.1.130.9894&rep=rep1&type=pdf, accessed March 28, 2019.
6. Personal communication to the authors, Sep. 2019.
7. K. F. Warnick, *Numerical Methods for Engineering: An Introduction Using MATLAB and Computational Electromagnetics Examples*, Edison, NJ: Scitech/IEE Press, 2011.
8. Teaching at Nottingham. *A Handbook of Values and Practice in a Research-Intensive University*, Spring 2011, p. 15.
9. K.T. Selvan, "A revisiting of scientific and philosophical perspectives on Maxwell's displacement current," *IEEE Antennas and Propagation Magazine*, vol. 51, no. 3, pp. 36-46, June 2009.
10. K.T. Selvan and S.R. Rengarajan, "*Teaching-in-context of Maxwell's displacement current: What do professors and students perceive?*," *2010 IEEE Antennas and Propagation Society International Symposium*, Toronto, Canada, July 11–17, 2010.
11. K. T. Selvan, "Engineering Education: Presentation of Maxwell's equations in historical perspective and the likely desirable outcomes," *IEEE Antennas and Propagation Magazine*, vol. 49, no. 5, pp. 155–160, Oct. 2007.
12. K.T. Selvan, L. Ellison, "Incorporation of historical context into teaching: Student perception at the University of Nottingham Malaysia Campus," *IEEE Antennas and Propagation Magazine*, vol. 49, no. 5, pp. 161–162, October 2007.

13. K. F. Warnick, "Engineering accreditation and assessment," *IEEE Antennas and Propagation Magazine*, vol. 54, no. 5, pp. 209–210, Oct. 2012.
14. P. C. Brown, H. L. Roediger III, and M. A. McDaniel, *Make It Stick: The Science of Successful Learning*, Cambridge, Massachusetts: Harvard University Press, 2014.
15. C. Furse, D. Ziegenfuss, and S. Bamberg, "Learning to teach in the flipped classroom," *IEEE Antennas and Propagation Society International Symposium*, pp. 910–911, 2014.
16. http://clt.curtin.edu.au/events/conferences/tlf/tlf1995/candy.html, accessed March 29, 2019.
17. http://users.umiacs.umd.edu/~resnik/why_teach.html, accessed March 29, 2019
18. K.T. Selvan and P.F. Wahid, *"Teaching electromagnetic theory: Beyond a focus on applications,"* *2015 IEEE 4th Asia-Pacific Conference on Antennas and Propagation*, Bali, June 30-July 3, 2015.
19. https://engineering-jobs.theiet.org/article/why-do-engineering-graduates-choose-non-engineering-careers-hint-it-s-not-because-of-the-financial-sector-/, accessed October 9, 2019.
20. https://www.bbc.com/news/education-14823042, accessed October 9, 2019
21. https://electronicsb2b.com/wp-content/uploads/2011/02/Part-2_Indian-Electronics-Industry.pdf, accessed October 9, 2019
22. K.T. Selvan, "Lessons learned from the IEEE AP-S Madras Chapter on electromagnetics education in India," *IEEE Antennas and Propagation Magazine*, vol. 63, no. 1, pp. 97–102, February 2021.
23. K.T. Selvan and K.F. Warnick, "A global vision for academic scholarship and professional development," *Forum for Electromagnetic Research Methodologies and Application Technologies (FERMAT)*, 2018. Available at https://www.e-fermat.org/education/selvan-edu-2018-03/, accessed March 29, 2019.

3   An Experiential Learning Approach in Electromagnetics Education

Hugo G. Espinosa, Jenny James, and *Nickolas Littman*
Griffith University, Australia

Thomas Fickenscher
Helmut Schmidt University, Germany

David V. Thiel
Griffith University, Australia

CONTENTS

3.1 INTRODUCTION

3.1.1 EXPERIENTIAL EDUCATION

John Dewey is considered one of the forefathers of experiential education [1]. He believed that education must include participation and cooperation and that people need to be in contact with groups of individuals, in order to expand their own personal ideas [2]. Not only did Dewey believe in creating a stronger sense of community through cooperative learning, he also believed that knowledge is created through the transformation of experience.

For Dewey, experience is always a dynamic two-way process. He describes it as a "transaction taking place between the individual and, what at the time, constitutes the environment" [3]. He elaborates on this two-way process, suggesting that all experience involves both *doing* and *undergoing*. Doing refers to the part of the experience where the learners are required to act upon something [4]. It is the purposeful engagement of the individual with the environment. Undergoing refers to the consequences of experience on the individual. If a connection is not made between these two parts of the experience, the extent to which we learn from the experience is limited [5].

Project-based learning (PBL) is one teaching method that involves both doing and undergoing. It requires students to investigate real-world questions or problems and then act on and communicate ideas. The value of PBL has been recognized for many years throughout the world and incorporated in many educational contexts, from elementary schools to universities. Research has shown that students with special needs, English-language learners, and talented and gifted students can all receive the same benefits from this approach [6], and that PBL approaches positively affect student performance and retention across elementary, secondary, and postsecondary levels [7].

Dewey proposed that learning from experience involves

1. observation of surrounding conditions;
2. knowledge of what has happened in similar situations in the past, a knowledge obtained partly by recollection and partly from the information, advice, and warning of those who have had a wider experience; and
3. judgment which puts together what is observed and what is recalled to see what they signify [4].

He argued that education is characterized by observations from an experience, reflecting on what was done and what was needed, and then forming conceptualizations based on those reflections and preexisting knowledge. Dewey also emphasized that learning was a cyclical process where each subsequent experience builds on past experiences.

Building on the work of Dewey, Kolb [8] argued learning is also a cycle that perpetuates more learning, and his theory of experiential learning elaborates the process by which students learn from their experience. Kolb's model (Figure 3.1) suggests four stages of experiential learning: *concrete experience, reflective observation, abstract conceptualization,* and *active experimentation* [8,9].

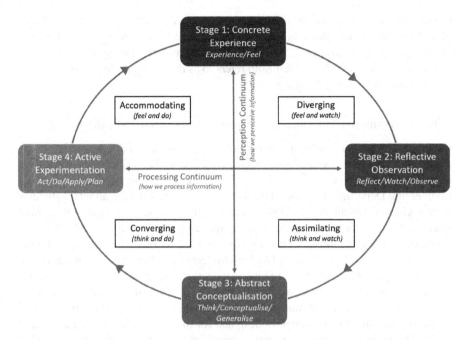

FIGURE 3.1 Kolb's theory of experiential learning: Learning cycle through four quadrants.

Kolb's cycle starts with a concrete experience, where the learner encounters a new experience or reinterprets an existing experience and considers what they can do, as well as what others have done previously. The next phase, reflective observation, involves taking time out from doing and focusing on what has been experienced and its implications considering past learning. In the third phase, abstract conceptualization, learners make sense of what has happened by interpreting the events and understanding the relationship between what they have done, with other past experiences, concepts, or principles they have encountered. During active experimentation, the last phase before the cycle begins again, the learner applies their learned concepts and theories to other realistic and practical activities and situations.

According to Kolb and Fry [10], the learner must progress through all stages of the learning cycle for one's learning to fully transform into one's understanding; however, the learner can enter the cycle at any one of the four stages. The student continues cycling through all four stages creating a learning spiral of ever-increasing complexity [11].

Kolb's learning theory [8] sets out four distinct learning styles based on this four-stage learning cycle. These learning styles are based on the ways people perceive and process information. The vertical axis in Figure 3.1 represents the *perception continuum* (how we perceive information) with one extreme being concrete experience (feeling) and the other being abstract conceptualization (thinking). The horizontal axis represents the *processing continuum* (how we process information) with one extreme being reflective observation (watching) and the other being active experimentation (doing) [12,13]. Kolb believed that it is only possible to operate on one end of a single continuum at any given time in the learning process (e.g., we are not able to simultaneously operate on the thinking and feeling ends of the perception continuum). Superimposing these lines on Kolb's learning cycle yields the four quadrants illustrating the four learning styles, which Kolb refers to as *diverging, assimilating, converging*, and *accommodating* (Figure 3.1).

3.1.2 EXPERIENTIAL LEARNING IN TRAINING PROFESSIONAL ENGINEERS

Experiential learning has been successfully applied in different engineering disciplines. In [14], Abdulwahed and Nagy explain how Kolb's learning cycle was implemented in the second-year Process Control Laboratory from the Chemical Engineering Department at Loughborough University, UK, where about 70 students were registered for the course. Through class presentations, pre-laboratory virtual work, hands-on practice on experimental rigs, and post-laboratory reflection sessions, students cycled through the four stages of the experiential process (see Figure 3.1). Holzer and Andruet [15], explain how students from Mechanical Engineering at Virginia Polytechnic Institute and State University, USA, cycled through the experiential learning process in their virtual laboratories for statics integrated with mechanics of materials. Interactive multimedia, traditional pencil-and-paper activities, and cooperative learning were strategies designed for the four stages of the experiential cycle. The literature, however, is rather scarce in relation to experiential learning applied to electromagnetics education, and this chapter may be one of the very few to report the learning cycle experiential process in an electromagnetics-based undergraduate electronic engineering course, providing examples of student work and feedback that verifies the effectiveness of Kolb's learning process.

Electromagnetics provides the basic foundation for electrical and electronic engineering systems and components in wireless communication, lightwave technology, radar, antennas, microelectronics, sensing technologies, and many more. Teaching advanced communications courses is an important part of undergraduate electrical, electronics, and microwave engineering education. As professional engineering recognition is a major driver of engineering education, students are required to graduate with Stage 1 competency levels, which include "fluent application of engineering techniques, tools and resources" and "application of established engineering methods to complex engineering problem solving" [www.engineersaustralia.org.au]. Despite the importance of advanced communications courses, an increasing disengagement in the field of electromagnetics has been observed among students [16]. In order to address this problem, platforms for the dissemination of new and trending strategies in electromagnetic education have been established. These include the IEEE MTT Education Forum [17], the special issue of the IEEE Antennas and Propagation Magazine: "From Engineering Electromagnetics to Electromagnetics Engineering: Teaching/training next generations" [18], and the convened session: Education in Electromagnetics, Antennas, and Microwave, from the European Conference on Antennas and Propagation (EuCAP 2020). This chapter is a revised and expanded version of a paper invited to such a session [19].

Furthermore, there is a worldwide awareness of the challenges faced in the methodology and practice of electromagnetics teaching and learning [20,21]. Several initiatives have been developed and implemented to move teaching from traditional face-to-face activities focusing on analytical or semi-analytical approaches toward practice-discovery sessions. The latter decouples, to a certain extent, the mathematics from the physical understanding, so that students will have an opportunity to become motivated and invested in electromagnetics before being required to undertake the challenge of studying the more complex, heavily mathematical constructs [22]. Some of these initiatives include PBL [23,24], problem-based learning [25], flipped classroom [26,27], experiential learning [19,28,29], competency-based learning [30], and other pedagogical approaches [31]. Such strategies intensively engage the student interactively in assigned pre-work based on conceptual questions and realistic examples in order to develop their problem-solving skills.

The different initiatives in electromagnetics education have sparked the development of several computer-aided design (CAD) tools, virtual laboratories, electromagnetic solvers, Matlab®-based computational tools, and even touch-based interactive tools through mobile technology, to support the curriculum and enhance the teaching and learning of different electromagnetic topics [32–38]. Such initiatives share the same objectives: to increase student skills and engagement, enhance student learning outcomes, improve knowledge retention, and create an ethical vision for improving society.

With the increase of engineering practice and internships before graduation, it is important to provide engineering students with the technical abilities and practical skills required in industry. Employers and Engineers Australia [www.engineersaustralia.org. au] require engineering graduates with not only technical knowledge, but also "hands-on" abilities such as design, measurement, data analysis, etc.

The following section describes the structure of the course 6303ENG: Advanced Communication Systems, including course learning outcomes, learning activities, and assessment. Furthermore, details are provided on how Kolb's theory of experiential learning has been applied to design activities (e.g., assignments) that provide students with self-driven experiential learning opportunities relevant to their own future careers. This course forms part of the four-year Bachelor in Electronic Engineering at Griffith University and is offered in the students' fourth year of study. Training in electromagnetics for communication engineers follows an annual progression of learning: Circuits and System (first year), Electromagnetic fields and Propagating Systems (second year), Linear Electromagnetics (third year), and Advances Communication Systems (fourth year).

Several examples of student work are presented as case studies. Each case study is discussed with results and visual aids from actual projects developed between the years 2017 and 2020. An assessment of the effectiveness of the experiential learning approach is reported by means of student feedback.

3.2 METHODOLOGY

This section provides details on the experiential learning cycle, the structure of the course used as a case study, and a description of the course assessment items, where students are required to develop and complete two experimental projects over a 12-week trimester. At the outset of each project, pairs of students are presented with a list of projects and, based on the brief outlines provided, are required to select a project to work on. Projects are unique to each two-person group with a sometimes obscure but practical industrial outcome designed to complement the lecture material. To succeed, students must continuously discuss their project strategies, measurements, and final applications with the teaching team throughout the trimester, and write a final report showing context and possible applications of the technology.

The course covers theoretical and practical aspects of microwave and UHF systems and components, and their applications in wireless communications systems. The teaching modules include microwave circuits and measurements, radio navigation and satellite communications, mobile wireless communications systems, radar, and smart antenna systems. Some of the modules are given by guest lecturers with expertise and experience in both academia and industry. The course is designed to acquaint students with technologies in the communications field. Students gain experience in the theory and practical implementation of a variety of communications technologies.

The course assumes an understanding of the application of Maxwell's equations to radiating systems. Students work in self-selected pairs to undertake research and development into two discrete and specified tasks with a list of available software and equipment. In each case, the outcome is a report on that task, which includes a summary of background knowledge, experimental and theoretical analysis of the task, results, and a reflective conclusion. The generic skills include task formulation, written and oral communication, data analysis, fabrication techniques, critical thinking and analysis, teamwork, and report construction. A major part of this course is self-driven experiential learning through independent projects and continuing engagement with the teaching team.

3.2.1 STRUCTURE OF THE COURSE

The course is offered in a 12-week period once a year; it consists of 24 hours of lectures (1 × 2 hrs/week) and 36 hours of laboratories (1 × 3 hrs/week). After completing the course, students should be able to:

- Assess reported outcomes using critical thinking and analysis based on engineering principles.
- Create an experimental plan, which includes verification of the experimental outcomes.

3.2.1.1 Course Assessment Items

The course assessment is composed of two practical assignments and one final examination as follows:

- Assignment 1: 20%
- Assignment 2: 20%
- Final written examination: 60%

Significant emphasis is placed on the two assignments. Given that a communications engineer requires both a strong theoretical understanding of the material and some practical and numerical modeling experience, an experiential learning approach was implemented for the assessment that allows students to apply their new knowledge and understanding in a problem-solving context.

For both Assignments 1 and 2, students have six weeks to select, conceive, design, implement, reflect, and report the outcomes of the project. Students are asked to provide a brief explanation of the theory, the experimental/numerical results, error calculation, and strong verification of the results through a comparison with other data. The assessment is based on the submission of a report with the following sections: Title, Abstract, Theory, Experimental methods, Results, Discussion, Reflection, and Conclusions. While students may work on the project in pairs, individual submissions of the report are required, reflecting the student's understanding and individual work. After submission of Assignment 1, students are then required to undertake an informal discussion, where they propose any improvements in their approach and strategies to be implemented in Assignment 2.

Assignments are marked out of 20 marks each. The marking scheme is based on the following scale:

- 15–20: excellent review of the problem, clear understanding of the measurements, and analysis of results.
- 10–14: some review of the problem, some understanding of the measurements, and analysis of results.
- 0–9: inadequate work completed, poor understanding of the problem and the practical techniques, poor analysis of results.

3.2.1.2 Laboratory Resources

Before entering the laboratory for the first time, students undertake a health and safety induction and complete a risk assessment. Each group is assigned an individual laboratory bench for the whole course. During the laboratory sessions, students have shared access to specialized equipment such as bench VNA – vector network analyzer (Anritsu MS46122A 1MHz–20GHz), handheld VNA (Keysight Field Fox N9923A 2MHz–6GHz), RF field strength analyzer (Protek 3290N 2–9GHz), signal generator (SL6000L 25MHz–6GHz), logarithmic RF detector (AD8319 100MHz–10.5GHz), digital lux meter (HoneyTek LX1010B), X-band waveguides, Gunn oscillators, frequency counters, and antenna test site. Software available includes NEC-Win Pro®, Matlab®, Altium Designer®, PUFF®, and Microwave Office®.

3.2.1.3 Assignment Topics

For Assignment 1, students receive a list of topics at the beginning of the course. They select a topic on a "first in" basis and that topic is no longer available to the rest of students. The list of Assignment 1 topics is carefully designed based on current technology trends, potential applications in industry, and relevance to lecture content. Every project is described in no more than three lines. The list provided to students includes around 30 available projects. Some examples include:

1. Scattering of a 3D printed conductor at X-band.
2. Effects of a C-slot on microstrip patch antenna bandwidth.
3. Buried monopole antenna performance using a finite ground plane at 2.8 GHz. The antenna will be buried in dry sand and both the antenna impedance and the radiation pattern will be measured.

4. Location uncertainties of a 500-MHz transmitter using two spaced receivers as a function of time and x–y position.
5. Cityscape scattering at X-band using a simple monopole as receiving antenna.
6. The effects of bends on a halfwave dipole antenna on a flexible paper substrate at 1 GHz. The antenna will be constructed using adhesive copper foil.
7. Microstrip T-match for a monopole antenna at 800 MHz on FR4 and plastic using a coplanar ground plane.
8. Underwater patch antenna properties at 2 GHz using a microstrip feed and a quarter-wave transformer.
9. X-band properties of concrete.
10. Open stub effective permittivity measurements on skin and the determination of anisotropy.
11. VLF surface impedance measurements and modeling over a long driven wire.
12. Wheeler cap measurements of the efficiency of a rectangular patch antenna.
13. 433-MHz field survey measurements in the vicinity of pipes and cables.
14. Surface measurements of the field distribution from a buried transmitter as a function of depth and polarization.
15. Radar scattering from an aluminum-coated golf ball.
16. Design, construction, and testing of a meanderline microstrip monopole with a coplanar ground plane at 2.5 GHz.
17. X-band horn antenna transmission system to detect water droplets by reflection and transmission.
18. Construction and testing of an air core, plastic balun formed from two small plastic cylinders with SMA terminations over a very wide frequency range.
19. Two-element meanderline phase-shifted microstrip antenna array.
20. Optical profiling of a laser beam in the visible spectrum using a multimode optical fiber and a translation stage.

As it can be seen in the previous list, topics include antenna design, link budget calculations, circuit design, and wave propagation. Depending on the topic, an emphasis can be given on the design, manufacture, simulation, or measurement techniques.

Once the student completes and submits Assignment 1 online through Blackboard,[1] the list of projects for Assignment 2 is made available. Students then have another six weeks for the completion and submission of the second report.

3.2.2 EXPERIENTIAL LEARNING CYCLE

Kolb's experiential learning model [8] was implemented, requiring students to complete a four-stage cycle of learning for each of the two assignments (one learning cycle per project). Although learners can usually enter the learning cycle at any one of the four stages, students were required to enter the cycle at the concrete experience stage for both assignments (see Figure 3.1).

During the concrete experience stage (stage 1), students are given a three-line description of the project from which they design a project plan based on fundamental theory provided during lecture time. In the reflective observation stage (stage 2), students observe the surrounding conditions and assess and reflect on the initial

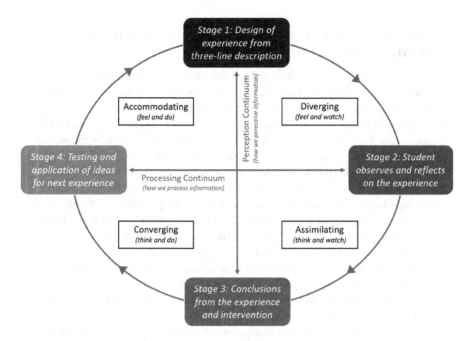

FIGURE 3.2 The four stages of Kolb's experiential learning cycle applied to the Advanced Communication Systems course.

outcomes of the experience. Project management, group work, and student-to-student interaction is important for a meaningful reflection; the questions "what worked?" and "what didn't work?" arise. In the abstract conceptualization stage (stage 3), students draw conclusions from knowledge obtained through the experience and published literature, and the advice and support from their peers and the teaching staff. Students can modify their new project strategy for Assignment 2. In the active experimentation stage (stage 4) students apply their revised project protocol and move back into the concrete experience stage (stage 1), to then continue through the remaining stages of the second cycle. Figure 3.2 shows an adaptation of Kolb's experiential learning cycle [8], where each of the four stages is described in terms of their application to the benchmarked course.

3.2.3 EXAMPLES OF STUDENT WORK

To evaluate the effectiveness and impact of the project-based experiential learning activities, seven assignments were selected as good examples of student work. The selection was based on the effective experimental design and measurement techniques, optimum use of resources, impact, and successful completion. A case study based on the work performed by a student in both assignments is demonstrated. Here, information about the student performance through the two-cycle process is presented; this includes the testing of their research hypothesis, the outcomes obtained in the first project that influenced the approach undertaken in the second project, and the role of the instructor in the student's achievements.

ASSIGNMENT 1   EFFECTS OF A C-SLOT ON MICROSTRIP PATCH ANTENNA BANDWIDTH

In this project, the student investigated the effects of the incorporation of a C-slot on the bandwidth of a rectangular microstrip patch antenna. A simple patch antenna was designed using the properties of the FR4 substrate used with a height of 1.6 mm and a relative permittivity of 4.6. Other design considerations were made using general rule-of-thumb techniques, such as extending the ground plane multiple times the thickness of the substrate beyond the antenna's footprint and the inclusion of a quarter-wave transformer to match the antenna's higher impedance calculated to be 255Ω to the lower characteristic impedance of the transmission line of 50Ω. The geometry of the slot was carefully considered, with the outer perimeter of the slot set to one half a wavelength of the desired frequency, with a thin slot width ensuring the slot was centrally positioned on the antenna and was opposite to the feedline to improve matching conditions. The experimental patch antenna (designed to operate at 3.9 GHz), Figure 3.3a, and the C-slot patch antenna, Figure 3.3b, were fabricated by laboratory technical staff using standard printed circuit board photolithographic techniques.

The bandwidth and central frequency were found by measuring the scattering parameters using a portable VNA, ensuring proper calibration, and the measurements were made without interference from nearby objects, which could affect the antenna's response. The S_{11} return loss measurements of both antennas, as seen in Figure 3.4, were determined. The student concluded that the inclusion of a C-slot resulted in a shift of the resonant frequency from 3.87 GHz (reference antenna designed for 3.9 GHz) to 3.53 GHz. This was consistent with the literature [39], which indicated that the inclusion of slots increases the length of the current paths within the antenna and, thus, would decrease the resonant frequency.

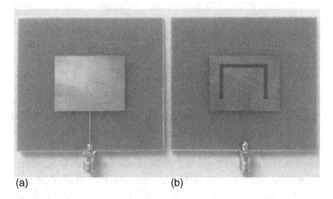

(a)   (b)

FIGURE 3.3 Proposed patch antennas with quarter-wave transformers and SMA connectors: (a) reference patch antenna and (b) C-slotted patch antenna.

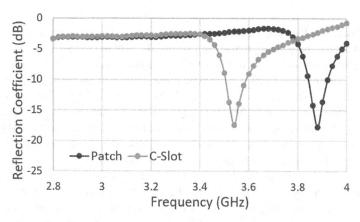

FIGURE 3.4 Return loss of the proposed patch antennas. A decrease in resonant frequency is evident in the slotted antenna compared to the reference antenna.

ASSIGNMENT 2 TWO-ELEMENT MEANDERLINE PHASE-SHIFTED MICROSTRIP ANTENNA ARRAY.

In this project, the student investigated the use of switched meanderlines as phase shifters in order to change the main beam direction of a two-element, microstrip patch antenna array. The patch antenna array was designed and fabricated using standard PCB manufacturing techniques. A quarter-wave transformer parallel feed network was used to evenly split the power between the identical antennas. Mitered bends were employed to reduce the impact of stray capacitances and, hence, impedance discontinuities in the meanderlines. PIN-diode switches along with DC-block capacitors of 0.001 µF were used to change the length of the meanderlines, which, when biased, would effectively short circuit a meanderline segment causing a phase shift between the antennas. Three diodes were used, with each diode representing a phase shift of 45°, 22.5°, and 11.25°, respectively. The meanderline networks for both antenna elements were symmetrical to ensure no phase shift was evident without a forward bias voltage of the diodes. It is also notable that the antennas and lines were adequately separated to ensure that the crosstalk and coupling was minimized, with measured crosstalk of around −112 dB for the largest meanderline segment and −32.1 dB for the smallest meanderline segment.

Prototypes of both a single-patch antenna combined with the proposed three-element switched meanderline phase shifter and two separate single-element switched meanderline phase shifters, shown in Figure 3.5a and Figure 3.5b, were fabricated to validate the phase-shifting technique as well as a two-element uniform array with meanderline phase-shifting segments, shown in Figure 3.5c. A calibrated VNA was used to measure the S_{11} and S_{21} scattering parameters of the prototype microstrip circuits, where a 1-Hz TTL signal supplied by a function generator was used to bias the appropriate diodes. Alligator clips were attached to the legs of the appropriate diode to allow for real-time observation of the main beam radiation direction due to the phase shifters. The return loss of the single-patch antenna showed a prominent null at the designed operating frequency of 2 GHz. The measured bandwidth was 4.7%, with several smaller nulls evident at higher frequencies, which are most likely due to the effects of the components on the PCB (see Figure 3.6). The prototype 11.25° and 22.5° phase shifter revealed shifts of 9.8° and 19.7°, respectively, at 2 GHz. This difference was mainly attributed to fabrication imperfections and parasitic capacitances and, although these phase shifts were not exactly as expected, they validated this phase-shifting technique.

The effects of the phase shifting on the array beamsteering were explored through the S_{12} scattering parameter, where the two-element phased array was used as the transmitting antenna and the single prototype patch antenna was used as the receiving antenna. A VNA was used to measure the frequency response of the path loss between the antennas fixed in place ensuring the receiving antenna was located in the far-field and broadside of the transmitting antenna array. Two phase shifts were measured with the appropriate diodes biased along with a control measurement with no phase shift. The impact of the

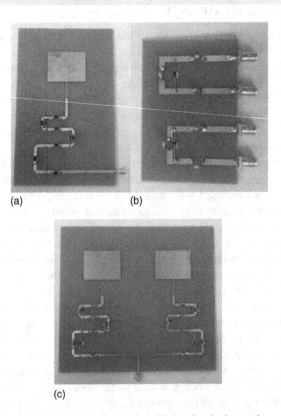

(a)   (b)

(c)

FIGURE 3.5   The designed microstrip circuits with (a) the single-patch antenna with phase-shifting meanderline, (b) 11.25° and 22.5° switched meanderline phase shifters, and (c) two-element antenna array with meanderline phase shifters and feed network.

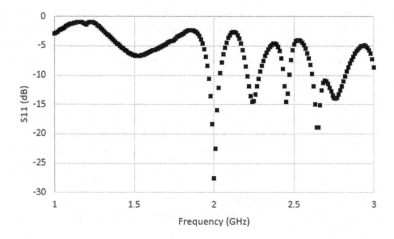

FIGURE 3.6 The return loss of the two-element patch antenna array with meanderline feeds.

beamsteering on the transmission loss reduces the magnitude of the received signal at resonance (2.0 GHz) from –41.02 dB to –44.46 dB and –47.89 dB, as the phase shift is increased from 0° to 11.25° and 22.5°, respectively. This is due to the main beam of the array steering away from the fixed receiver located at broadside, consequently reducing its received power. The student concluded that the meandered phase shifters were a viable beamsteering option.

After completion of both assignments, the student was asked to undertake the following survey to evaluate the effectiveness of experiential learning as an assessment methodology:

Q1. Did you find the assessment process easy to follow?
Answer: The assessment process was easy to follow as the projects were clearly outlined.

Q2. Did you apply the learned process from the 1st assignment toward the 2nd assignment?
Answer: The first assignment allowed for reflection and more preparedness before the second assignment, with more experience and a greater understanding compared to the first assignment.

Q3. What outcomes did you obtain in Assignment 1 that influenced the approach that you undertook in the cycle process of Assignment 2?
Answer: As a response to Assignment 1, prototypes were created in Assignment 2 to validate the technique along with additional design and planning to improve the project. The inclusion of the prototype allowed for better measurement and confirmation that the phase shifting was working as intended. In addition, it allowed for a greater understanding and improve-

*ment of the quality of the assignment, which also tied into reflecting on
the first for a better planning and approach of the second. Such approach
was different to the approach undertaken in the first project as it allowed
for better quantification of the effects of the meanders, which were measured directly in the prototypes.*

Q4.   What activities did you perform on each of the four stages of the learning cycle?
*Answer: In stage 1, the project was chosen and discussed with the instructors,
and a basic design and plan for the experiment was conducted. Expected
outcomes and theoretical modeling/calculations were also performed.*

In stage 2, the experiment was conducted, and results were gathered.

*In stage 3, the experimental and theoretical results were compared,
and a conclusion was drawn.*

*In stage 4, reflection and improvements for the next project were
undertaken.*

Q5.   What was the role of the instructor or how did you perceive the instructor's
role in your achievements of the two assignments?
*Answer: Instructors acted as additional groupmates, they were involved in discus-
sions and provided support and feedback, allowing for the discussion of
ideas, which were invaluable to the project.*

3.2.3.1   Compilation of Student Assignments

A compilation of further examples of student work based on the assignment topics
described in section 3.2.1.3 is included as follows.

Example 3.1: X-Band properties of concrete

In this project, the student investigated the electromagnetic properties of concrete
at X-band as a key step toward improving communication systems in urban areas.
The properties of a strong concrete mix were analyzed.

A concrete block of size $10 \times 10 \times 35$ cm^3 was used for the experiment. One
corner of the block had a copper rod set into the concrete along one of the short
axes. A pair of horn antennas was placed on either side in contact with the concrete (Figure 3.7). One of the horn antennas was connected to a radiofrequency
(RF) signal generator and was used as the transmitter. An identical horn (receiver)
was connected to a logarithmic detector.

The experiment was conducted by manually stepping the generator from 6 GHz
to 12 GHz in steps of 100 MHz. This procedure was performed for three different
lateral positions of the pair of horn antennas, one at each end of the concrete
block (2.5 cm from the edge), and another positioned in the center of the block.

Figure 3.8 shows the received voltage from the logarithmic detector as a function of frequency, for the three different antenna positions. The student used the
frequency response to calculate the relative permittivity and conductivity of the
concrete.

The student concluded that the concrete material is a low loss dielectric in the
frequency band 6–10 GHz. The relative permittivity and intrinsic impedance of
the concrete were calculated to be 3.69 and 196 Ω, respectively. This showed
good agreement with published work on various concrete mixes [40].

FIGURE 3.7   Isometric view of the horn transmission experiment through concrete.

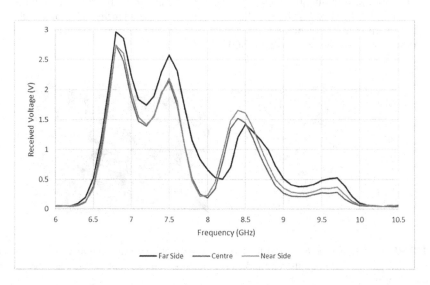

FIGURE 3.8   Frequency response measurements for each of the three sets of measurements along the sample (left, center, and right). The resonance peaks were used to calculate the electromagnetic properties of the concrete.

Example 3.2: Cityscape scattering at X-band using a simple monopole as receiving antenna

In this project, the student investigated the effect of reflective surfaces that scatter, absorb, and reflect radio waves, and its impact on radio communication systems in urban environments.

A simple collinear, half-wave monopole antenna was constructed using a semi-rigid coaxial cable with an unshielded quarter-wave extension and tuned using a ferrite bead on the outer conductor. A model city was constructed at a scale of approximately 1:2500. Black cardstock of 200 gsm was used to create structures, and aluminum foil was used to provide RF reflectance. A number of different-sized boxes were constructed and lined with the aluminum foil to represent the buildings (Figure 3.9). Holes were drilled through the ground surface in an approximate grid for antenna placement. The model city (an approximate representation of several actual city blocks in Brisbane, Australia) was placed 15 cm away from the horn antenna to ensure operation in the far-field. The 10.5 GHz Gunn Oscillator illuminated the model via a horn antenna with the beam directed along the length of the simulated streets (scaled from 4 MHz in the true situation).

Measurements were initially taken at each point of the grid without the buildings, by inserting the antenna (connected to a detector and DC voltmeter) vertically through each hole. After the buildings were placed, the difference in field strength was determined at each point on the grid. A number of blackspots and hotspots were located. The resulting intensity heatmap is shown in Figure 3.10.

This is in line with the Rayleigh and Rician fading characteristics and multipath ray tracing covered in the lecture material on mobile communications. The major blackspot had an attenuation loss of −18.2 dB compared to the open-plane measurements. The strongest hotspot had an increased field intensity value of 5.8 dB. Given the number of different reflective surfaces, this level of constructive and destructive interference was expected.

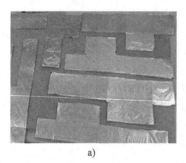

a)   b)

FIGURE 3.9 Experimental setup for the cityscape scattering: (a) aluminum foil-wrapped cardstock and (b) Gunn oscillator, model city, antenna, detector (not visible), and voltmeter (not visible).

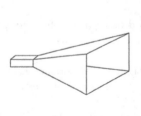

FIGURE 3.10 Attenuation heatmap overlayed on cityscape. The source was located on the center left of the image. Red signifies an area of low signal strength and green signifies an area of stronger signal strength.

Example 3.3: The effects of bending a half-wave dipole antenna on a flexible paper substrate

In this project, the student explored the effect of bending a half-wave dipole antenna on a flexible substrate, by comparing the simulated and measured return loss, S_{11}. The antenna was fabricated by adhering two strips of copper tape totaling 15.5 cm in length to a 29.7 cm paper substrate. An SMA connector was fixed to the paper substrate where the inner conductor was soldered to one arm of the antenna with the outer conductor soldered to the other arm, through a transformer balun as shown in Figure 3.11.

NEC-Win Pro was used to model the antenna, where the antenna parameters, as well as varying the bending radii, were input to determine the return loss. A VNA was used to measure the return loss, where a baseline flat measurement was made by placing the antenna on a wooden block, before elevating the center of the antenna causing the central bend. The simulated return loss shown in Figure 3.12a and measured return loss shown in Figure 3.12b were in good agreement, with both cases showing a decrease in return loss magnitude at resonance as the antenna was bent.

The simulated resonant frequency (Figure 3.12a) for the baseline measurement where the antenna was not bent, was 930 MHz, while the corresponding measured resonant frequency was significantly lower at around 700 MHz. This was most likely due to the cardboard and the wooden board that the antenna was resting upon, which would have increased the antenna's effective length due to its relative permittivity. Similarly, the other measured results for the resonant frequency at around 840MHz (Figure 3.12b) were lower than the simulated results at around 930MHz, which, again, are most likely due to the effect of the wooden block along with nonideal measurement conditions.

FIGURE 3.11   The fabricated flexible half wave-dipole antenna.

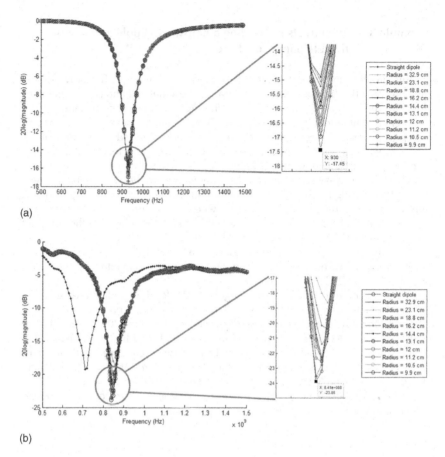

FIGURE 3.12 Return loss of the dipole antenna with arc bend obtained by (a) numerical field simulation and (b) measurements.

Example 3.4: Radar scattering from an aluminum coated golf ball

In this project, the student investigated the scattering characteristics of an aluminum coated golf ball, where the parameters target range, bistatic angle, and polarization were explored. The impact of target range on the backscattered signal was measured with two horn antennas placed next to each other (quasi-monostatic setup), directed at the golf ball, so that the backscattered signal was received by the receiving antenna, with varying distance between the ball and the antennas. A CW radar with log diode detector receiver was used to measure the received power, where the measured DC voltage is proportional to the microwave power, as shown in Figure 3.13. Two distinct range regions are apparent: the first linear region from 4 to 14 cm, where the power is maximum and it follows a linear decay (dark gray), and a second region, where the power no longer follows a linear decaying relationship, and the signal is indistinguishable from noise (light gray).

The bistatic radar scattering was measured using two horn antennas with the receiving antenna rotated in 10° increments, as seen in Figure 3.14, with the golf ball fixed in place on a foam stand on the axis of the rotating table. Three different

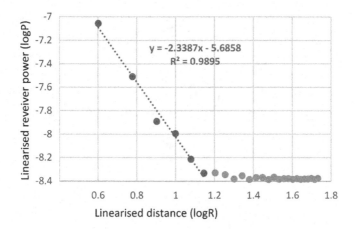

FIGURE 3.13 Linearized measured backscatter from an aluminum coated gold ball with two distinct regions and a fitting function in the linear dropping region.

FIGURE 3.14 Experimental setup for bistatic scattering measurements. The setup shown was used for the baseline measurements, after which the golf ball was placed in the center of the rotation table and held constant while the receiver was rotated.

distances from the transmitter were measured, with and without the ball, to create a baseline. The receiver voltage magnitude decreased with distance in line with the previous experiment as seen in Figure 3.15. For all cases, the detected voltage decreased significantly between 90° and 270° (backscatter region), with some side lobes evident in the smaller distances. Interference and scattering from the golf ball is evident when compared with the baseline measurements at the same distance, with significant decreases in power between 80° and 280°.

Finally, the polarimetric bistatic radar scattering of the golf ball was investigated using a similar setup as for the bistatic scattering measurement. The measured voltage received for both the E and H planes displayed similar values, as seen in Figure 3.16, which was to be expected as the golf ball was assumed to have a response similar to a perfect sphere with the scattering not dependent on polarization. The measurement results for the radar scattering for the cross-polarization were below the noise level.

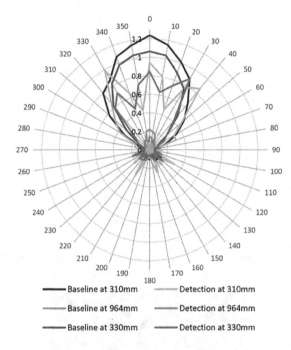

FIGURE 3.15   Measured bistatic scattering at three different distances.

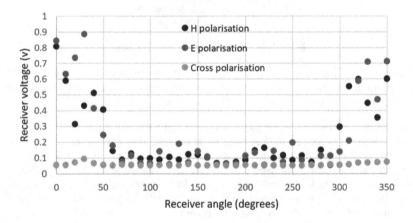

FIGURE 3.16 Measured receiver voltage for E-polarization, H-polarization, and cross-polarization.

Example 3.5: VLF surface impedance measurements and modeling over a long driven wire

In this project, the student explored the surface impedance of a radio field using a driven wire along the earth's surface. Very low frequency (VLF) surface impedance can provide information about the subsurface structure of the earth, where anomalies in the conductivity can be identified. The surface impedance can be explored through the simulations measurement of the electric and magnetic fields surrounding the long driven wire using the Thiel Surface Impedance Meter (TSIM), an electronic hand-held instrument developed at Griffith University, and designed to measure impedances of earth at depths of up to 100 m [41,42].

A 20-m-long insulated wire was placed on the earth's surface and grounded at one end by an aluminum stake to create a circuit. A battery-powered low-frequency oscillator was used to drive the wire. The E- and H-field magnitudes were measured every 0.5m for 30 m at an angle of 45° with respect to the wire as illustrated in Figure 3.17.

Three sets of measurements were taken, one with the transmitter switched off to determine the background noise across the non-driven wire, another with the transmitter on and the wire driven with a low amplitude, and another with the transmitter on and the wire driven with a high amplitude. Typical baseline measurements ranged between 35 and 45 V/m for the E-field, and between 15 and 25 A/m for the H-field using the NWC 19.8 kHz transmitter[2].

The E-field results, seen in Figure 3.18a, highlight the differences between the measured results and the theoretical inverse relationship between distance and magnitude. The theoretical maximum is when the distance between the driven wire and TSIM is the least, at 10 m along the line of measurements; however, the

FIGURE 3.17 Site location of the field measurements, with the black line representing the line of measurements and the gray line representing the driven wire, with the transmitter on the northern end. Inset: TSIM showing the gray driven wire crossing visible immediately below the coil on the left side of the instrument.

measured results show a maximum further along the line. The measured results show an increased field strength at the point directly above the wire, with significant interferences that could be attributed to either the subsurface anatomies or, more likely, the driven wire transmitter itself, as no abnormities were evident in the background noise measurements at this distance. The TSIM was not equidistant from the transmitter for the entirety of the measurement line and, therefore, may have had an effect on the fields due to this angle. This effect was not accounted for in the theoretical calculations. Distinct interference nulls can also be seen either side of the main null, where the measured amplitude for both cases is comparable to the background noise level. This was attributed to lateral waves caused by subsurface conductors reradiating the incident electromagnetic waves. Another noticeable difference in the E-field results is that the measured results do not decay to zero with distance from the wire as in the calculated fields.

The measured H-field results, in Figure 3.18b, showed good agreement for the higher amplitude case, where its maximum was very close to the theoretical maximum at 10 m. Noticeable fluctuations and deviations from the theoretical results are visible on the side of the peak; however, the measured fields return to the background noise level.

The experimentally-calculated surface impedance, in Figure 3.19, reflected the same peak displacement as seen in the measured results for the E-field, which is expected as these fields were used to calculate the surface impedance as it is defined as the ratio of horizontal electric field to the horizontal magnetic field.

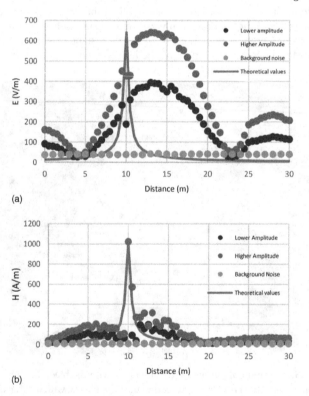

(a)

(b)

FIGURE 3.18 Comparison of measured (a) E-field and (b) H-field results for a wire driven at two different amplitudes with theoretical values and the background noise.

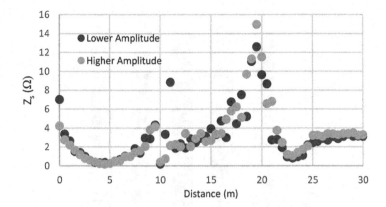

FIGURE 3.19 Surface impedance calculated from the experimental results, showing a distinct peak at 20 m for both amplitudes.

Throughout these assessments, the role of the instructor changes from the "sage on the stage" to the "guide by the side." Functioning more as a mentor or coach, the instructor helps students progress through the problems by questioning, probing, encouraging critical reflection, and providing ongoing feedback on what students have learned from their experiences.

3.2.4 STUDENT AND ALUMNI REFLECTIONS AND FEEDBACK

To assess the effectiveness of the course and the newly implemented experiential learning approach, anonymous student feedback was collected through a general qualitative and quantitative survey containing the following questions:

Q1. This course was well-organized.
Q2. The assessment was clear and fair.
Q3. I received helpful feedback on my assessment work.
Q4. The course engaged me in learning.
Q5. The teaching team was effective in helping me to learn.

The questions were scored using a 5-point Likert-type response scale, from 1 (strongly disagree) to 5 (strongly agree). The response rate was 65%, with 39 out of 60 undergraduate students submitting completed surveys. Figure 3.20 summarizes their responses. The 5-point scale was converted to a total score out of 100.

According to the feedback, students found the course engaging (average score of 92/100), highlighting the support and intervention from the teaching team during the projects.

The students found that the experiential learning approach to the assessment provided them with opportunities to tackle realistic communications engineering problems similar to those they would be asked to solve in the real world. They thought the theory complemented the practice very well and, although they found it challenging to only receive a brief (maximum of three lines) project description, they felt more

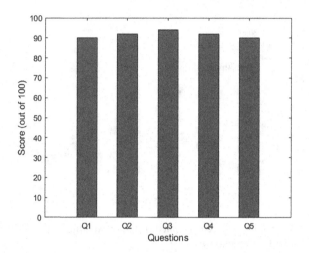

FIGURE 3.20 Anonymous survey response from undergraduate students. Questions 1 to 5 scored out of 100.

rewarded and satisfied at the end of the assessment, than if they had received step-by-step guided laboratory experiences.

Students performed overall much higher in Assignment 2, with an average of 65/100 for the first project and 87/100 for the second project. This supports the efficacy of the active experimentation stage of the learning cycle as students were able to test their research hypothesis, building into the creation of new and improved strategies for the following experience.

The following excerpts from student feedback provide some insight into their learning experiences:

> *The labs were incredibly useful and valuable as they gave us insights [in]to practical communication systems.*

> *The project assignments were a great opportunity to get hands-on with the theor[ies] taught and understand their application in a practical environment.*

> *I really enjoyed the labs, the projects were intensive and challenging, but they kept me very engaged with the course.*

Furthermore, one graduate who completed the course in 2017, and is now working as a Software Systems Engineer in Brisbane, Queensland, Australia, stated that she uses the experiential learning cycle for all her project work. After submitting her project for review and receiving feedback from her colleagues she then documents what she has learned and how the project can be improved for the next cycle. She also claims that, when possible, she uses data and results from previous projects to forecast expected results for future projects. This aligns with the experiential learning theories proposed by both Dewey and Kolb, that learning is a cyclical process where each subsequent experience builds on past experiences.

3.3 CONCLUSIONS

This chapter described how a project-based experiential learning approach was implemented into an undergraduate electromagnetics course at Griffith University over the past four years. Applying Kolb's experiential learning cycle, learning activities were designed that required students to integrate their theoretical and practical knowledge to investigate real-world problems and challenges. The students received a list of assessment topics, in the form of a series of brief (i.e., no more than three-line) project descriptions, from which they could choose two topics that were personally meaningful to them. Allowing students to select their own topics increases students' motivation and allows them to become active participants and co-creators of their education. Students are then required to circulate twice through Kolb's learning cycle (once for each project), working collaboratively with their peers to solve the project tasks, with the instructor facilitating and guiding the learning process. Throughout these assessment tasks, students are limited only by the laboratory facilities and resources provided. This type of assessment practice allows students to acquire transferrable skills that can be applied in the workplace upon graduation, including critical thinking and problem-solving, creativity and innovation, communication, collaboration, metacognition, and motivation.

NOTES

1 Blackboard is a virtual learning environment and learning management system developed by Blackboard Inc. (https://www.blackboard.com/). It is the learning management system used at Griffith University, Australia.
2 The transmitter is located at the Naval Communication Station Harold E. Holt in the North West Cape (NWC) peninsula of Western Australia. The station provides a VLF signal of 19.8 kHz with a transmission power of 1 MW.

REFERENCES

1. K. Warren, M. Sakofs and J. S. Hunt, *"The theory of experiential education,"* Dubuque, Iowa, USA: Kendall Hunt Pub Co; 3rd Ed., 1995.
2. S. D. Wurdinger, *"Philosophical issues in adventure education,"* Dubuque, Iowa, USA: Kendall Hunt Publishing; 3rd Ed., 1997.
3. J. Dewey, *"Experience and education,"* Indianapolis, IN, USA: Kappa Delta Pi, 1938.
4. J. Dewey, "Democracy and education," J. A. Boydston (Ed.), *The middle works*, vol. 9. Southern, Illinois, USA: University Press, 2008.
5. A. R. English, "Dewey on thinking in education," M. Peters (Ed.), *Encyclopedia of Educational Philosophy and Theory*, Singapore: Springer, 2016.
6. K. A. Hovey and S. L. Ferguson, "Teacher perspectives and experiences: Using project-based learning with exceptional and diverse students," *Curriculum & Teaching Dialogue*, vol. 16, no, 1/2, pp. 77–90, 2014.
7. V. L. Holmes and Y. Hwang, "Exploring the effects of project-based learning in secondary mathematics education," *The Journal of Educational Research*, vol. 109, no. 5, pp. 449–463, 2016.
8. D. A. Kolb, *"Experimental learning: Experience as the source of learning and development,"* Upper Saddle River, New Jersey, USA: Pearson 2nd Ed., 2014.

9. C. Manolis, D. J. Burns, R. Assudani and R. Chinta, "Assessing experiential learning styles: A methodological reconstruction and validation of the Kolb Learning Style Inventory," *Learning and Individual Differences*, vol. 23, pp. 44–52, 2013.

10. D. A. Kolb and R. Fry, "Towards an applied theory of experiential learning," C. L. Cooper (Ed.), *Theories of group processes*. NY, Wiley, 1975.

11. L. H. Lewis and C. J. Williams, L. Jackson and R. S. Caffarella, "Experiential learning: A new approach," *New Directions for Adult and Continuing Education*, vol. 62, pp. 5–16, 1994.

12. A. Konak, T. K. Clark and M. Nasereddin, "Using Kolb's experiential learning cycle to improve student learning in virtual computer laboratories," *Computers & Education*, vol. 72, pp. 11–22, 2014.

13. J. E. Sharp, J. N. Harb and R. E. Terry, "Combining Kolb learning styles and writing to learn in engineering classes," *The Research Journal for Engineering Education*, vol. 86, no. 2, pp. 93–101, 2013.

14. M. Abdulwahed and Z. K. Nagy, "Applying Kolb's experiential learning cycle for laboratory education," *Journal of Engineering Education*, vol. 98, no. 3, pp. 283–294, 2009.

15. S. M. Holzer and R. H. Andruet, "Experiential learning in mechanics with multimedia," *International Journal of Engineering Education*, vol. 16, no. 5, pp. 372–384, 2000.

16. H. Roussel and M. Helier, "Difficulties in teaching electromagnetism: an eight year experience at Pierre and Marie Curie Universtiy," *Advanced Electromagnetics*, vol 1. no. 1. pp. 65–69, 2012.

17. K. Wu, "Challenges and opportunities for education in RF/microwaves", *IEEE Microwave Magazine*, vol. 17, no. 7, pp 10–12, July 2016.

18. S. V. Hum and L. Sevgi, "From engineering electromagnetics to electromagnetic engineering: Teaching/training next generations," *IEEE Antennas and Propagation Magazine*, vol. 62, no. 2, pp. 12–13, 2020.

19. H. G. Espinosa, T. Fickenscher, N. Littman and D. V. Thiel, *"Teaching wireless communications courses: An experiential learning approach,"* EuCAP *2020 (on-line), 14th European Conf. Antennas and Prop*, Copenhagen, Denmark, March 2020.

20. B. M. Notaros, *"Electromagnetics education and its future and challenges,"* EuCAP *2019, 13th European Conf. Antennas and Prop*, Krakov, Poland, June 2019.

21. A. Sihvola, *"Challenges for electromagnetics teaching and education,"* FERMAT, *Forum for Electromagnetic Research Methods and Application Technologies*, Guangzhou, August 2014.

22. N. Anderson and M. Mina, "A new approach in teaching electromagnetism: How to teach EM to all levels from freshman to graduate and advanced-level students," *Proceedings of American Society for Engineering Education*, pp. 1–17, 2003.

23. P. Crilly, *"An innovative approach to teaching an undergraduate electromagnetics, antennas and propagation course,"* Proc. of the 2014 Zone 1 Conference of the American Society for Engineering Education, 2014.

24. A. Nieves, J. Urbina, T. Kane, S. Huang and D. Penaloza, "Work in progress for developing project-based experiential learning of engineering electromagnetics," *ASEE Annual Conference & Exposition*, 2019.

25. C. Mias, "Electronic problem based learning of electromagnetics through software development," *Computer Applications in Engineering Education*, vol. 16, no. 1, pp. 12–20, 2008.

26. M. B. Cohen and A. Zajic, *"Revitalizing electromagnetics education with the flipped classroom,"* USNC-URSI Radio Science Meeting, Vancouver, BC, Canada, July 2015.

27. C. M. Furse and D. H. Ziegenfuss, "A busy professor's guide to sanely flipping your classroom," *IEEE Antennas and Propagation Magazine*, vol. 62, no. 2, pp. 31–42, 2020.

28. M. J. Schroeder, A. Kottsick, J. Lee, M. Newell, J. Purcell and R. M. Nelson, "Experiential learning of electromagnetic concepts through designing, building and calibrating a broad-spectrum suite of sensors in a capstone course," *International Journal of Electrical Engineering Education*, vol. 46, no. 2, pp. 198–210, 2009.

29. J. Ernst, "Impact of experiential learning on cognitive outcome in technology and engineering teacher preparation," *Journal of Technology Education*, vol. 24, no. 2, pp. 31–40, 2013.

30. G. Artner and C. F. Mecklenbrauker, "The competence-oriented teaching of antennas, propagation, and wireless communications, *IEEE Antennas and Propagation Magazine*, vol. 62, no. 2, pp. 43–49, 2020.

31. J. Leppavirta, H. Kettunen and A. Sihvola, "Complex problem exercises in developing engineering students' conceptual and procedural knowledge of electromagnetics," *IEEE Transactions on Education*, vol. 54, no. 1, pp. 63–66, 2010.

32. B. M. Notaros, R. McCullough, S. B. Manic and A. Maciejewski, "Computer assisted learning of electromagnetics through MATLAB programming of electromagnetic fields in the creativity thread of an integrated approach to electrical engineering education," *Computer Applications in Engineering Education*, vol. 27, no. 2, pp. 271–287, 2019.

33. L. C. Trintinalia, "Simulation tool for the visualization of EM wave reflection and refraction," *IEEE Antennas and Propagation Magazine*, vol. 55, no. 1, pp. 203–211, 2013.

34. H. G. Espinosa and D. V. Thiel, "MATLAB-Based interactive tool for teaching electromagnetics," *IEEE Antennas and Propagation Magazine*, vol. 59, no. 5, pp. 140–146, 2017.

35. L. Sevgi, *"From engineering electromagnetics to electromagnetic engineering: Teaching/training next generations,"* EuCAP 2019, 13th European Conf. Antennas and Prop, Krakov, Poland, June 2019.

36. L. Sevgi, "Electromagnetic modeling and simulation: Challenges in validation, verification and calibration," *IEEE Trans on EMC*, vol. 56, no 4, pp. 750–758, 2014.

37. A. Polo, M. Salucci, A. Gelmini, G. Gottardi, G. Oliversi, P. Rossa and A. Massa, *"Advanced teaching in EM – Towards the integration of theoretical skills and applicative/industrial skills"*, EuCAP 2019, 13th European Conf. Antennas and Prop, Krakov, Poland, June 2019.

38. E. L. Tan and D. Y. Heh, "Mobile teaching and learning of coupled-line structures," *IEEE Antennas and Propagation Magazine*, vol. 62, no. 2, pp. 62–69, 2020.

39. S. Bhardwaj and Y. R. Samii, "A comparative study of C-shaped, E-shaped, and U-slotted patch antennas," *Microwave and Optical Technology Letters*, vol. 54, no. 7, pp. 1746–1756, 2012.

40. R. H. Haddad and I. L. Al-Qadi, "Characterisation of Portland cement concrete using electromagnetic waves over the microwave frequencies," *Cement and Concrete Research*, vol. 28, no. 10, pp. 1379–1391, 1998.

41. H. G. Espinosa and D. V. Thiel, "Efficient forward modelling using the self-consistent impedance method for electromagnetic surface impedance," *Exploration Geophysics*, vol. 45, no. 3, pp. 201–207, 2014.

42. G. T. Mogensen, H. G. Espinosa and D. V. Thiel, "Surface impedance mapping using sferics," *IEEE Transactions on Geoscience and Remote Sensing*, vol. 52, no. 4, 2074–2080, 2013.

4 Teaching and Learning Electromagnetics through MATLAB® Programming of Electromagnetic Fields

Branislav M. Notaroš, Ryan McCullough,
Sanja B. Manić, and *Anthony A. Maciejewski*
Colorado State University, USA

CONTENTS

4.1 INTRODUCTION

4.1.1 Overview of Pedagogical Approach

Supported by a five-year RED (REvolutionizing engineering and computer science Departments) grant from the National Science Foundation (NSF), a diverse team of educators at Colorado State University (CSU) is implementing a new approach to teaching and learning that reimagines the roles of the faculty and moves away from the traditional course-centric structure [1]. As described in the *IEEE Access* article "A Holistic Approach to Transforming Undergraduate Electrical Engineering Education" [2], our pedagogical model builds on the concept of "nanocourses" to facilitate knowledge integration (KI), a learning model grounded in education pedagogy and supported by research. The approach blurs the lines between courses because the faculty take a systems view of the curriculum to identify the fundamental technical concepts of an electrical and computer engineering (ECE) education, independent of courses. These concepts are then rearranged and organized into cohesive learning studio modules (LSMs) to lay the groundwork for real-world applications. Each LSM is self-contained and addresses one anchoring concept and a set of subtopics in a given core competency area [1,2]. Although a departure from the traditional course structure, LSMs still provide a path for students to learn all the intended topics in a rigorous manner.

Aiming to connect abstract concepts to the real world of engineering, KI activities are then created to put learning in context and illustrate the societal relevance of engineering knowledge [1,2]. Serving as a mechanism for helping students grasp the commonality and correlations between core concepts across the curriculum, KI activities show students how LSM fundamentals are integrated to form the building blocks of a complex piece of ubiquitous technology.

4.1.2 Creativity in the Technical Core of the Curriculum

While our five-year RED project spans the entire undergraduate program, special attention is given to the technical core, or junior year, of the ECE curriculum. In addition to instilling deep technical knowledge of the discipline, faculty are working in partnership to interweave the following threads throughout the curriculum to help students develop skills that will allow them to thrive as engineers [2]:

- Creativity thread – integrates research, design, and optimization
- Foundations thread – illustrates why math matters in the world of engineering
- Professional formation thread – emphasizes professional skills deemed important by industry

This paper dives into the creativity thread of the project, sharing details about how the department is using learning technology tools, in this case MATLAB, to enrich and assess learning.

4.1.3 CHALLENGES OF ELECTROMAGNETICS INSTRUCTION AND LEARNING

For the RED project, technical core material encompasses signals and systems, electromagnetics, and electronics. This paper focuses on the role and importance of electromagnetics in the undergraduate curriculum, and the challenges of its instruction and learning. It has been noted by several researchers [3–6] that students have an opinion that introductory electromagnetics is a difficult subject, and instructors also find it a difficult subject to teach. Indeed, while electromagnetic theory or theory of electromagnetic fields and waves is a fundamental underpinning of technical education, it is often perceived as the most challenging and demanding course in the electrical engineering (EE) curriculum. The material is extremely abstract and mathematically rigorous, and students find it difficult to grasp, which is not unique to any particular school/department, country, or geographical region.

Researchers attribute the difficulty to three main factors: (1) the use of vector mathematics, which some students can view as rather abstract; (2) introductory classes frequently only cover very idealized situations that do not have true physical applications; and (3) realistic electromagnetics examples in a laboratory setting are difficult to create. These three issues create various problems derived from commonly researched factors in engineering student attrition and achievement. Dinov, Sanchez, and Christou [7] also point out that classes that rely heavily on mathematics tend to miss other learning modalities such as visual and active learning. If the difficulty related to visual and active learning is not addressed, it becomes an issue with the instructor/student learning styles mismatch, as has been reported in multiple studies [8–12]. In addition, if the realistic examples and true physical applications are not presented in a way that is both rigorous and relevant, students tend to lose motivation [13].

4.1.4 UNDERSTANDING STUDENT LEARNING STYLES

Learning styles are characteristic preferences for alternative ways of taking in, and processing, information. The theory of learning styles has been studied for decades, beginning with Kolb's learning styles model in 1984. There are different variations on learning styles, but Felder and Silverman's Index of Learning Styles (ILS) [9] is commonly used in engineering education. They categorize five pairs of complementary learning styles: sensing and intuitive; visual and auditory; inductive and deductive; active and reflective; and sequential and global.

- Sensing and Intuitive: Sensing learners deal with the outside world through observing and gathering data through the senses, while intuitive learners indirectly perceive information through speculation and imagination.
- Visual and Auditory: Visual learners use sights, pictures, and diagrams, whereas auditory learners prefer sounds and text.
- Inductive and Deductive: Inductive learners prefer a reasoning progression from observations, measurements, data, etc., to generalities, i.e., governing rules, laws, and theories. Deductive learners prefer the opposite progression of inductive learning, going from the governing ideas to explaining new observations.

- Active and Reflective: Active learning takes place when experimentation involves discussions, explanations, or testing in the outside world. Reflective learners, meanwhile, prefer to go through these processes more introspectively. Kinesthetic learning is another commonly acknowledged modality of learning, where learners prefer exploring through touching and interacting, but Felder and Silverman consider this to be part of the active learning style.

- Sequential and Global: Sequential learners prefer the standard layout of class content where the concepts are introduced and learned systematically. Conversely, global learners prefer to learn in fits and starts, needing to see how all the pieces fit together before they can make sense of any small part.

4.1.5 IMPORTANCE OF ADAPTING INSTRUCTIONAL METHODS TO MOTIVATE LEARNERS

Although there is some debate on the effectiveness of matching instructional methods to a student's assessed learning style, there have been significant studies showing that a mismatch in learning styles leads to decreased performance and a higher attrition rate [8–12]. In Seymour and Hewitt's study "Talking about Leaving" [14], the data showed that grade distributions of students who leave technical curricula are essentially the same as the distributions of those who continue. Their findings revealed that some of the higher performing students leave because of dissatisfaction with their instruction. This fact was also recorded by Bernold, Spurlin, and Anson [10] in their three-year, 1000+-student study, where they systematically tested the learning style preferences and behaviors of the students and tracked their successes, failures, and paths throughout three years of the engineering program. They found that students who had the greatest mismatch in learning style preference versus the standard teaching style preference (lectures) had the largest attrition rate and the poorest grade point average (GPA), even when the comparison was done through a Scholastic Assessment Test (SAT) mathematics covariation. However, as Litzinger, Lee, Wise, and Felder [12] note, students with any learning style preference have the potential to succeed at any endeavor. Learning styles and modalities are simply preferences that dictate the ways in which people feel the most comfortable learning. Felder and Brent [11] note that how much a student learns in a class is governed in part by that student's native ability and prior preparation but also by the compatibility of the student's attributes as a learner and the instructor's teaching style.

A large part of student performance is tied to their motivation. College students, like all adults, have an issue motivating themselves to study material if it is not applicable to their lives or their future professions [15,16]. Ulaby and Hauck [6] observed that students learning in the standard lecture-style environment tend to doubt the usefulness of learning the subject of electromagnetics as they cannot see how it can be applied directly to other parts of the curriculum and how it will benefit them once they graduate. The cognitive science-based study "How People Learn" [16] recommends a model to mitigate this issue, which calls for instructional activities to focus on the most important principles and methods of a subject while building on the learner's current knowledge and conceptions. The activity should also utilize techniques known to promote skill development, conceptual understanding, and meta-cognitive awareness rather than simple factual recall.

4.1.6 BACKGROUND ON COMPUTER-ASSISTED LEARNING AND PROGRAMMING IN TECHNICAL EDUCATION

Computer software has been used to aid student learning in electromagnetics for quite some time [17–36]; although this use has expanded in recent years, it is not widespread yet and is not universally implemented. As Mias [31] notes, computer-assisted learning significantly improves the teaching of electromagnetics. He goes on to discuss some of the currently utilized methods of incorporating computer-assisted learning including educational graphical interfaces for electromagnetic field visualization, the use of spreadsheet programs in solving electromagnetic problems, the use of university and/or industry-developed computational electromagnetics software to solve real-life problems and gain an insight on electromagnetic field phenomena, and virtual laboratories. These solutions, although they can be shown to improve teaching, lack one specific task: programming. Mias [31] and Read [17] discuss the advantages of utilizing programming in electromagnetics instruction. They both note that the benefit of writing computer code is that it increases the students' need to understand fundamental field concepts of the problem involved and the need to be able to obtain the conventional analytical solutions of simple problems. Programming, as well as the other methods of computer-assisted learning, also brings in the possibility for design work, a core aspect of the creativity thread within the RED project, in a way that was not possible through traditional methods.

As another example of the effectiveness of computer-assisted learning in engineering/science education, generally, we look to the work by Dinov, Sanchez, and Christou [7]. In instructing several statistics courses of various levels, these authors experimented with a statistics online library provided by the NSF called Statistics Online Computational Resource (SOCR). While integrating this resource into their classrooms and post-work did not result in statistically significant gains in comparison to the traditional classroom, they did receive much more positive feedback for the course through a qualitative survey, and the retention rate of the supplemented classes was significantly higher. This shows that even if students do not learn more through programming, which is contrary to what Mias and Read suggest, they still enjoy the content matter more and are more willing to stick with the class.

The use of programming in the current EE curriculum generally follows the same trend in that students are frequently only required to take a single programming course at the beginning of their program. These courses, which are generally offered through the university's Computer Science (CS) department, teach students topics more related to CS, such as sorting and searching, rather than topics that are better suited to EE, such as matrix manipulation [3]. The students are then expected to use these rudimentary, unrelated programming skills to later program concepts such as vector algebra and calculus, multivariable functions, three-dimensional (3D) spatial visualization, numerical integration, optimization, and finite-difference and finite-element methods. This can be a struggle for many students without a stronger background in matrix manipulation and calculus-related programming.

4.1.7 UTILIZING MATLAB TO DEEPEN LEARNING AND INSPIRE CREATIVITY

To adapt our instructional methods to motivate students, and address the learning considerations outlined previously, we have chosen MATLAB® (by MathWorks, Inc.) as the learning technology and modeling software language for the creativity thread of the RED project. Evidence shows that students find MATLAB easy to use [17], and it is considered an important tool that ECE students and future engineers need to use effectively. By having a single software platform that is consistently used throughout the curriculum, students gain proficiency with the programming environment, allowing them to shift their focus from learning programming to understanding the technical content of what they are programming.

This paper presents inclusion of computer-assisted MATLAB-based instruction and learning in the electromagnetics course and LSMs of the RED project, where the students are implementing the core LSM concepts they learned into a "virtual electromagnetics testbed" using MATLAB, as part of the creativity thread. The students are taught "hands-on" electromagnetics through MATLAB-based electromagnetics tutorials and assignments of exercises and projects in MATLAB. To enable this, the lead author of this paper has developed a unique and comprehensive collection of approximately 400 MATLAB computer exercises, problems, and projects, covering and reinforcing practically all important theoretical concepts, methodologies, and problem-solving techniques in electromagnetic fields and waves [37], as a modern tool for learning electromagnetics via computer-mediated exploration and inquiry. These tutorials, exercises, and codes are designed to maximally exploit the technological and pedagogical power of MATLAB software as a general learning technology. Similar to the study done by Dinov, Sanchez, and Christou [7], the results of this work are qualitatively analyzed through a survey given to the students that recorded their feedback about the integration of MATLAB in their coursework. A preliminary report on this work appears in a conference proceedings [38].

When MATLAB has previously been chosen as a platform for computer-assisted learning [27–36], the instructors frequently have created a graphical user interface (GUI) or some similar interface that the students interact with as opposed to writing their own programs. As discussed above, programming forces students to think about the problem logically and to pay attention to the details, while still having a big picture of the problem at hand. Simply letting a GUI do the calculations and imaging the results is not as conducive to the learning process as the active process of programming electromagnetics.

Note that the "creativity" term within this study should be considered and appreciated in a broad sense. The "creativity thread" of our RED project is an umbrella for a variety of research, analysis, and design-related activities that students perform individually and in teams within the individual courses and through projects at different stages in the curriculum. Through this thread, we are attempting to inspire and enhance not only students' creativity but also a number of other, related, abilities and skills essential for their preparation for the real world of engineering. Likewise, the link between the "creativity" and the inclusion of computer-assisted MATLAB-based instruction and learning in the ECE electromagnetics course and LSMs as presented and discussed in this work should be understood broadly, in multiple ways. Foremost, the "creativity" term linked to the use of MATLAB in electromagnetics classes here comes from the context of this presented implementation, namely, as part of the creativity thread of the RED project, and practically all references to "creativity" in this paper are within this context. In addition, MATLAB exercises can help the students develop a stronger intuition and a deeper understanding of electromagnetic field theory, examples, and problems, which is a prerequisite for reaching other (higher) categories of learning, including analyzing, evaluating, and creating. Through MATLAB-based computer-mediated exploration and inquiry, students can also invoke and enhance their analytical, evaluative, and creative thinking about and dealing with electromagnetic fields by exploring various "what ifs" and bridging the fundamental aspects and applications without being overwhelmed with abstract and often overly complicated and dry mathematics of analytical solutions. Furthermore, most programming tasks in support of scientific and engineering computing are, arguably, creative to some extent, and so is the proposed MATLAB programming of electromagnetics performed by students, as opposed to a passive computer demonstration. Finally, our MATLAB problems are designed and assigned to help the students develop and enhance their MATLAB skills irrespective of the electromagnetics context, which can then be utilized in other courses in the curriculum, as well as in research activities and projects, and there even more closely and directly tied with the innovation and creativity – for example, within design work.

4.1.8   NOVELTY AND BROADER IMPACTS OF PROPOSED MATLAB PROGRAMMING OF ELECTROMAGNETICS IN THE CREATIVITY THREAD

Overall, the principal novelty of our approach is the introduction of MATLAB programming of electromagnetics performed by students, as opposed to passive instructional methods. Our approach has a twofold goal for students as learners: (i) gaining and solidifying their knowledge of fundamentals of electromagnetic fields and (ii) building and enhancing their understanding and command of MATLAB syntax, functionality, and programming in the framework of electromagnetics. Our approach attempts to capitalize on a win-win combination of the pedagogical benefits of MATLAB-based electromagnetics education and students' computer-related skills and interests.

Another benefit of our approach is that it enhances students' MATLAB skills, which are used throughout the creativity thread of our integrated approach to ECE education, as well as in KI activities that span multiple courses across the curriculum. In addition, some of the most important pedagogical features of MATLAB related to the instruction and comprehension of electromagnetic fields and other ECE topics are its abilities to manipulate and visualize vectors and spatially distributed physical quantities, numerically solve problems, and use symbolic programming to reinforce analytical solutions.

This paper shares examples of the MATLAB assignments we delivered as part of the creativity thread. We also demonstrate how these assignments engage a broad range of learning styles, to include those students who do not thrive in a lecture-style learning environment, such as visual, active, global, and inductive learners. We also illustrate how our approach increases students' motivation to learn, their attitudes toward the subject, and appreciation of the practical relevance of the material. Finally, our paper outlines how our work helps the students develop and improve the understanding and command of MATLAB use and programming, within and beyond the electromagnetics context.

4.2   METHODS AND IMPLEMENTATION

4.2.1   INTRODUCING MATLAB PROGRAMMING OF ELECTROMAGNETIC FIELDS IN AN ELECTROMAGNETICS COURSE

MATLAB-based instruction and learning was introduced in ECE 341, the Electromagnetic Fields I course in the ECE Department at CSU during the fall semester of 2016. This is the first course of a mandatory sequence of two electromagnetics courses covering static and low-frequency (quasistatic) electric, magnetic, and electromagnetic fields (commonly referred to as a classical fields course) with the second course in the sequence dealing with generation and propagation of unbounded and guided electromagnetic waves and transmission lines, namely, a waves course. The fields course was integrated, through the RED project, with signals/systems and electronics courses running in the same semester. The five LSMs of the course are as follows: LSM1 Electrostatic Field in Free Space; LSM2 Electrostatic Field in Material Media; LSM3 Steady Electric Currents; LSM4 Magnetostatic Field; and LSM5 Low-Frequency Electromagnetic Field.

In creativity class sessions, students were given MATLAB tutorials/lectures, with ample discussions of approaches, programming strategies, MATLAB formalities, and alternatives. These were followed by comprehensive and rather challenging multi-week homework assignments consisting of MATLAB problems and projects in electromagnetic fields. In addition to the creativity class lectures, each MATLAB assignment included a number of tutorial exercises with detailed solutions combined with listings of MATLAB code. Even these exercises were assigned and graded, as their completion required creation of MATLAB programs from the provided portions of code, actual execution of the program, and generation and presentation of the results. Normally, all new concepts, approaches, and techniques in MATLAB programming as applied to electromagnetic fields were covered in tutorials, to provide students with additional guidance for completing similar exercises on their own.

Some specific technical and pedagogical features of these MATLAB exercises, projects, and codes include vector field computation, visualization of spatially distributed scalar and vector quantities, symbolic and numerical programming, e.g., symbolic and numerical integration, and solutions to nonlinear problems. There are many other features that may be suitable and implementable into a fields class [37], that were beyond our particular implementation in this study, and will be part of our future work. Another large set of features would apply to the waves course or the waves portion of the electromagnetics course in a single-course scenario [37].

The MATLAB exercises and projects for the electromagnetic fields course were created and chosen by the instructor of the course to best support the electromagnetics LSMs and the RED project as a whole [37]. The instructor also authored the textbook used for the course [39], and, as such, was able to choose MATLAB exercises that fluently supported the traditional course content. As a part of the assignments, the students were expected to take the core LSM concepts they learned in class and implement and explore them in the form of computer exercises utilizing MATLAB.

Essentially, the students are learning MATLAB in the context of electromagnetics and learning electromagnetics in the context of MATLAB. Moreover, in our opinion and experience, including programming actively challenges and involves the student, providing additional, prolonged benefits of learning as compared to a passive computer demonstration. This is consistent with the observations described by both Hoole [3] and Hoburg [40]: The mere use of computers does not enhance learning skills; it is when students develop their own programs or general-purpose software that their curiosity and creativity is aroused and they learn the most.

The observations described by Hoole and Hoburg above also fit well with the educational theory of learning styles, discussed in the Introduction. Although simply using computers to model a system can help the visual learners to see a visual representation of the electric and magnetic fields, it does not help active learners engage with the material. Also, because students do not have to consider the fundamental field concepts to simply use a computer, the students do not gain a full understanding of the origins behind the fields, a key aspect of electromagnetism. By presenting the overlying phenomena as a question/problem and using a programming approach to modeling fields and forces, students are required to build the solution from a set of underlying concepts, an aspect of inductive learning. This quality is a key factor in understanding material. As Felder and Silverman [9] state in their paper "Learning and Teaching Styles in Engineering Education," the benefits claimed for inductive learning include increased academic achievement, enhanced reasoning skills, longer retention of information, improved ability to apply principles, and increased capability for inventive thought. The learning style advantages of programming versus normal simulation software do not end there. The ability to manipulate the program and explore various "what ifs" and applications while still considering the fundamental aspects at play are a key component of learning electromagnetism through a global perspective. Therefore, through using programming to model electric and magnetic fields in given situations, the learning styles less frequently reached by typical lecture-style teaching and standard simulation software (active, global, and inductive) are all achieved through our approach.

As explained in the Introduction, learning styles were not our only consideration when designing and choosing the MATLAB exercises/problems; we also considered the relevance of the problems in terms of actual engineering design and their relationship to current material covered in other courses. Some of the problems simply cannot be related to the students' everyday lives and engineering designs because many of the basic concepts are not strictly applied; however, when there was an application, we tried to include it. For example, one problem required the students to use a GUI calculator that calculates the capacitance of a microstrip line and other common transmission lines and capacitor structures. Some of the more abstract relationships were also addressed in the creativity section. For instance, as most of the programming dealt with the visualization of electric and magnetic fields and forces, the ideas of electromagnetic interference (EMI) were addressed and discussed to stress the importance of knowing the field structures and forms to be able to predict where EMI and various types of capacitive, conductive, inductive, radiative, etc., couplings could be an issue.

By writing and executing their own MATLAB programs to solve problems and generate results presented in figures, diagrams, movies, and animations, students gain a stronger intuition and deeper understanding of electromagnetic fields, one of the most difficult subjects in EE, primarily because it is extremely abstract. Simultaneously, these diverse MATLAB projects and exercises help students to gain comprehensive operational proficiency in concepts and techniques of MATLAB use and programming. This knowledge and skill can then be applied effectively in other areas of study, including other courses in the curriculum, most notably in the signals/systems and electronics courses in the junior year, which is the focus of the RED project.

In addition, as Hoole [3] suggests, computer modeling and programming in the context of engineering education allows the students to not only gain an understanding of content, but to confront and appreciate the reality that design is an iterative modeling and analysis process. To help students address and recognize this process and its benefits, some MATLAB exercises/problems required students to create programs to solve problems that they already had to solve "by hand" for homework or that were done "conventionally" in class. Some other problems asked the students to implement the fundamental concepts and equations from the course in MATLAB programs, carry out calculations and visualizations using their programs, and draw and report relevant conclusions and results. Another class of problems directly addressed real-world practical applications of electric and magnetic fields discussed in the course. All problems were designed and assigned to help the students develop and enhance their MATLAB skills, irrespective of the electromagnetics context.

4.2.2 ILLUSTRATIVE EXAMPLES FROM CREATIVITY THREAD MATLAB ASSIGNMENTS IN ELECTROMAGNETICS CLASSES

Example Problems 1 and 2 – given in Table 4.1 and Figures 4.1 and 4.2 – show how some of the MATLAB problems encourage students to think about how small changes to the MATLAB program can lead to vastly different results. The results from these examples also show the power of MATLAB to visualize vectors, vector algebra, and 3D spatially distributed quantities. As mentioned above, this clearly addresses the visual learning style that is normally ignored in a traditional lecture. In terms of MATLAB programming skill development, Problem 1, for example, teaches the students to perform vector numerical integration, and to visualize vector fields by means of arrows in 2D space using the MATLAB function "quiver," having

TABLE 4.1

Example Problems 1-4 from the Creativity Thread MATLAB Assignments in the Electromagnetics (Fields) Class

Problem #	MATLAB Problem Statement
1	MATLAB function "quiver" is used for visualization of field vectors in 2D space. Input data are coordinates of nodes in a mesh in a Cartesian coordinate system and intensities of field components at the nodes. Implement "quiver" to visualize the electric field distribution due to a uniform straight line charge of finite length l and total charge Q placed along the x-axis in free space. Although the analytical solution is available in this case, given by [39] $$\mathbf{E} = \frac{Q}{4\pi\varepsilon_0 ld}\left[\left(\cos\theta_2 - \cos\theta_1\right)\hat{\mathbf{x}} + \left(\sin\theta_2 - \sin\theta_1\right)\hat{\mathbf{y}}\right]$$ for the situation shown in Figure 4.1, the electric field vector at each node of the mesh should be computed by vector numerical integration of elementary fields due to equivalent point charges along the line representing short segments into which the line is subdivided. With such an integration (superposition) procedure, this MATLAB program may be applicable, with appropriate modifications, to many similar and more complex charge distributions, where the analytical expression for electric field components is not available or is difficult to find. Output from the MATLAB code is shown in Figure 4.2.
2	Repeat the previous MATLAB Exercise but for three equal point charges Q residing at vertices of an equilateral triangle of side a in free space.
3	Write a program in MATLAB that uses previously created programs and calculates and plots the electric force on a point charge due to N other point charges in free space. The input to the program consists of N, coordinates of charge points, and charges $Q_1, Q_2, \ldots, Q_N$, as well as coordinates and charge of the point charge on which the force is evaluated. Then use this code to plot the force on one of three equal point charges at vertices of an arbitrary triangle. Figure 4.3 shows an output from the MATLAB code.
4	Write a MATLAB program that displays the distribution of the electric potential due to an electric dipole (Figure 4.4) with a moment $\mathbf{p} = Q d\mathbf{z}$ located at the origin of a spherical coordinate system. As output, the program provides two plots in the plane defined by $y = 0$ in the associated Cartesian coordinate system: one representing the potential by means of MATLAB function "pcolor" (that uses color to visualize the third dimension) and the other showing equipotential lines with the help of MATLAB function "contour." Depicted in Figure 4.5 is an output from the MATLAB code.

previously defining a 2D mesh of field points using the MATLAB function "mesh-grid." None of these MATLAB tasks are simple and all have many different applications.

Moreover, in Example Problem 1, vector integration is carried out based on spatial subdivision of the integration domain, approximation of charges on integration segments as equivalent point charges, and computation of the Cartesian vector components of the total electric field at each node of the field mesh as a sum of all elemental field components – in MATLAB. This exactly coincides with the analytical procedure and thought flow taught in regular class sessions for obtaining the fields due to given continuous charge distributions by vector summation of the fields

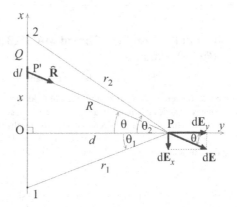

FIGURE 4.1 Geometry for Example Problem 1 (from MATLAB assignment 1) in Table 4.1 [39].

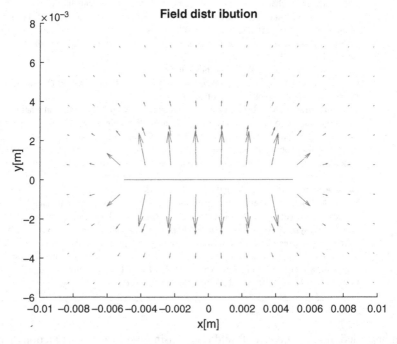

FIGURE 4.2 Graphical output from the MATLAB code for Example Problem 1 in Table 4.1, for the following input: Line length (Figure 4.1) in cm: 1; Total charge in nC: 1.

contributed by the numerous equivalent point charges making up the charge distribution. So students are now "instructing" a computer to evaluate the electric field due to a spatial charge distribution by vector integration in a computer program in essentially the same way as they were instructed to do so analytically in fields classes. There perhaps is no better way for students to acquire and embrace the principle of superposition, which is a consequence of the linearity of the electromagnetic system

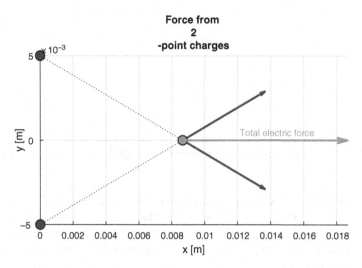

FIGURE 4.3 Graphical output from the MATLAB code for Example Problem 3 (from MATLAB assignment 1) in Table 4.1, for an adopted position of charges.

in this and most other cases and is one of the most important principles in electromagnetic (and ECE in general) analysis, computation, and design, than to "teach" the computer how to do it based on fields analytics from classes. This is key in developing students' problem-solving (analytical and computational) skills for true 2D and 3D vector problems, and essentially any complex problem. While there is no universally optimal algorithm for solving complex problems, it is wise to use superposition whenever possible – i.e., to break up a complex problem into simpler ones, and then add up (integrate) their solutions to get the solution to the original problem [39]. These fundamental concepts of linearity and superposition are reinforced across multiple ECE subjects, especially in the signals/systems and electronics courses within the RED KI activities.

Example Problem 3, presented in Table 4.1 and Figure 4.3, also shows an example of a vector visualization question/problem and MATLAB programming task given to the students, but now extends the visualization from 2D to 3D space by using the MATLAB function "quiver3." In addition, this problem helps students to recognize and appreciate the vector electric forces, both the components and the resultant, due to point electric charges. The resultant force on a charge is computed again using the principle of superposition, as the vector sum of partial forces due to other charges in the system. Although this and similar problems are only ideal and do not have immediate counterparts in reality, they are still useful as the students have to consider the fundamental concepts to program the system, and the end result helps them to visualize field effects, a vital capability when addressing some real-life engineering problems.

Example Problem 4 (Table 4.1 and Figures 4.4 and 4.5) helps students visualize spatial patterns and changes in a scalar variable using the MATLAB functions "pcolor" (with color continuously representing the function value as the third dimension above a 2D cut of 3D space) and "contour" (with colored lines representing cuts

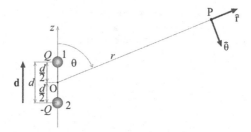

FIGURE 4.4 Geometry for Example Problem 4 (from MATLAB assignment 1) in Table 4.1 [39].

in a plane of surfaces having the same value of a 3D spatially dependent quantity at all points – equipotential, V = const, surfaces in this case). It also develops and solidifies the comprehension and mastery of an important concept of an electric dipole (Figure 4.4) and the electric potential (and consequently the electric field vector) that it produces in 3D space.

Examples of MATLAB problems that require the students to compare a MATLAB solution to the "by hand" solution are included as Example Problems 5 and 6, given in Table 4.2 and Figures 4.6–4.8. Generally, students are often specifically asked to redo some of the conventional computational problems they had for homework (typically, Problems from the book [39]) or that were done in class (typically, Examples from the book [39]), namely, either the same or similar problems, now using MATLAB to experience firsthand the power and utility of numerical and symbolic analysis and computation. Solving the problems and studying the topics both analytically and using MATLAB is extremely beneficial, as it significantly reinforces the fundamental concepts required for both approaches and ensures students are addressing the active, global, and inductive learning styles. Note that Example Problems 5 and 6, given in the Creativity Thread MATLAB assignments in the fields class, as well as the Related Classical Problems in Table 4.2, done in class or for homework, are all "physical," realistic, practical, interesting, nontrivial problems, and not purely formulaic (plug-and-chug) or purely "mathematical," dry, formal problems. They are very challenging to address, analyze, and solve both analytically and in MATLAB, and thus very beneficial for students' learning in both contexts. In addition, students are challenged to implement decision making and equation setting/solving processes that they are carrying out analytically when solving problems classically into a general algorithm and software so the computer can provide the result for any set of input values, and not just for the ones specified in a particular problem.

Figures 4.9–4.12 present additional example results from the MATLAB assignments illustrating the power of MATLAB computation and visualization in electromagnetics (fields). Namely, Figure 4.9 shows the results of a question/problem that required students to address the important concept of boundary conditions for a dielectric–dielectric boundary. The programming aspect enforces the fundamental concepts, utilizing a detailed tutorial for a similar, preceding, MATLAB problem, while the result encourages a visual learning perspective.

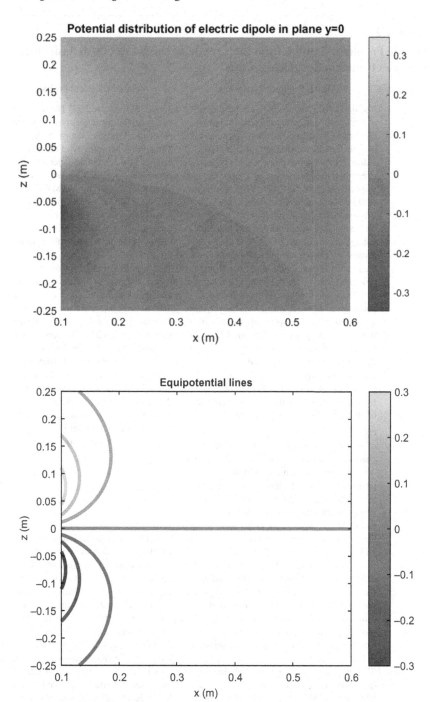

FIGURE 4.5 Graphical output from the MATLAB code for Example Problem 4 (from MATLAB assignment 1) in Table 4.1 (Figure 4.4), using MATLAB functions "pcolor" (top panel) and "contour" (bottom panel).

TABLE 4.2
Example Problems 5 and 6 from the Creativity Thread MATLAB Assignments in the Electromagnetics (Fields) Class

Problem #	MATLAB Problem Statement	Related Classical Problem
5	This MATLAB Exercise analyzes breakdown in a spherical capacitor with a multilayer dielectric. Consider a spherical capacitor with N concentric dielectric layers. The inner radius, relative permittivity, and the dielectric strength of the ith layer are a_i, ε_{ri}, and E_{cri}, respectively ($i = 1, 2, ..., N$). The inner radius of the outer conductor of the capacitor is b. Write a MATLAB program to find which dielectric layer would break down first after a voltage of critical value is applied across the capacitor electrodes and to compute the breakdown voltage of the capacitor. Test the program for $N = 3$, $a_1 = 2.5$ cm, $a_2 = 5$ cm, $a_3 = 7$ cm, and $b = 9$ cm, if the dielectrics constituting layers 1, 2, and 3 are polystyrene, quartz, and silicon, respectively (use GUIs from MATLAB Exercises I.16 and I.17 to get the values of material parameters). HINT: The data input is realized as in MATLAB Exercise I.3. We implement a generalization of Eqs. (2.241) and (2.242) from the book [39] to the case of an arbitrary number (N) of layers, and, instead of only two charges $Q_{cr}^{(1)}$ and $Q_{cr}^{(2)}$, we now have an array of such charges. For finding the minimum charge in the array, we use MATLAB function "min," which also returns as output the position (index) of the minimum value in the array. Then the corresponding breakdown voltage is computed based on Eqs. (2.240) and (2.241) from the book [39].	(Book [39] Example 2.32) The dielectric of a spherical capacitor consists of two concentric layers, as shown in Figure 4.6. The relative permittivity of the inner layer is $\varepsilon_{r1} = 2.5$ and its dielectric strength $E_{cr1} = 50$ MV/m. For the outer layer, $\varepsilon_{r2} = 5$ and $E_{cr2} = 30$ MV/m. Electrode radii are a = 3 cm and c = 8 cm, and the radius of the boundary surface between the layers is $b = 5$ cm. Calculate the breakdown voltage of the capacitor. (Book [39] Problem 2.75) Consider the spherical capacitor with two concentric dielectric layers in Figure 4.6 and assume that the inner dielectric layer is made from mica ($\varepsilon_{r1} = 5.4$), whereas the outer layer is oil ($\varepsilon_{r2} = 2.3$). The geometrical parameters are $a = 2$ cm, $b = 8$ cm, and $c = 16$ cm. The dielectric strengths for mica and oil are $E_{cr1} = 200$ MV/m and $E_{cr2} = 15$ MV/m, respectively. The oil is then drained from the capacitor. Find the breakdown voltage of the capacitor in (a) the first state (outer layer is oil) and (b) the second state (outer layer is air).
6	Consider a simple nonlinear magnetic circuit with an air gap, shown in Figure 4.7, and redo Example 5.13 from the book [39] but now in MATLAB. Write a MATLAB code that plots the magnetization curve of the ferromagnetic core and the load line of the circuit (in the same graph), and computes numerically and plots the operating point of the circuit. Compare the result to the analytical solution in Example 5.13. HINT: Use function "magCurveSolution" (from MATLAB Exercise II.15) and see the tutorials to MATLAB Exercises II.17 and II.4. Figure 4.8 shows the graphical presentation of the MATLAB solution.	(Book [39] Example 5.13; also book [39] Problem 5.21) Consider a magnetic circuit consisting of a thin toroidal ferromagnetic core with a coil and an air gap shown in Figure 4.7(a). The coil has $N = 1000$ turns of wire with the total resistance $R = 50$ Ω. The length of the ferromagnetic portion of the circuit is $l = 1$ m, the thickness (width) of the gap is $l_0 = 4$ mm, the cross-sectional area of the toroid is $S = S_0 = 5$ cm^2, and the emf of the generator in the coil circuit is E = 200 V. The idealized initial magnetization curve of the material is given in Figure 4.7(b). Find the magnetic field intensities in the core and in the air gap.

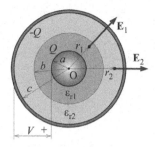

FIGURE 4.6 Geometry and material properties for Example Problem 5 (from MATLAB assignment 2) in Table 4.2 [39].

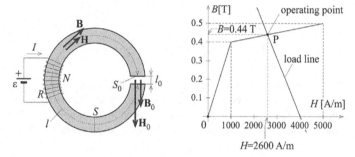

FIGURE 4.7 Geometry and material properties for Example Problem 6 (from MATLAB assignment 2) in Table 4.2; also shown is the analytical solution at the operating point for the nonlinear magnetic circuit [39].

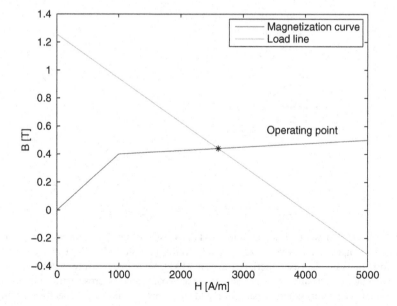

FIGURE 4.8 Graphical output from the MATLAB code for Example Problem 6 in Table 4.2, showing the MATLAB solution of the nonlinear magnetic circuit in Figure 4.7.

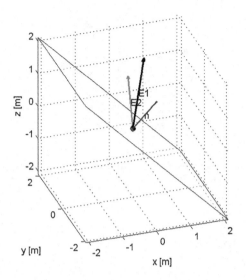

FIGURE 4.9 An example from MATLAB assignment 1 illustrating the power of MATLAB computation and visualization in electromagnetics (fields): MATLAB computation and visualization of dielectric-dielectric vector field boundary conditions for an arbitrarily positioned (oblique) boundary plane between media 1 and 2 with arbitrary given permittivities (dielectric constants).

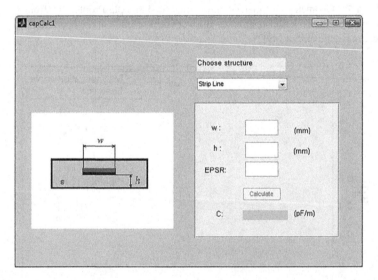

FIGURE 4.10 Another example from MATLAB assignment 1 illustrating the power of MATLAB computation and visualization in electromagnetics (fields): Capacitance calculator in the form of a graphical user interface (GUI) in MATLAB to interactively calculate and show the capacitance or per-unit-length capacitance of a coaxial cable, microstrip transmission line, parallel-plate capacitor, spherical capacitor, and strip transmission line, respectively, with the names of structures appearing in a pop-up menu.

Electron travel in a uniform magnetic field

FIGURE 4.11   An example from MATLAB assignment 2 illustrating the power of MATLAB computation and visualization in electromagnetics (fields): Snapshot of a movie in MATLAB that traces the path traveled by an electron as it enters a uniform time-constant magnetic field.

Figure 4.10 depicts an example where the programming was actually done for the students to create a GUI. This GUI, however, is very useful as it helps the students to understand and utilize the capacitance "produced" by different capacitors and transmission line geometries. This addresses the students' need for problems relevant in their own lives. In addition, interested students were able to explore and play with the MATLAB script (source code) for the GUI and learn firsthand, or at least get a rough idea, how such real-life complex GUIs are designed, developed, and coded in MATLAB.

Figure 4.11 portrays an example problem that also addresses a main concept in electromagnetism while having a very specialized application to real life in areas like particle accelerators. The programming behind the result will once again force students to fully consider the fundamental concepts and any special issues due to the logic requirements of a computer.

Finally, Figure 4.12 addresses the common visualization issue when time-harmonic alternating current exists in a straight-wire conductor, and, even more difficult, the visualization of the nearby electric field it produces. Because these concepts are hard to grasp and visualize, the results show them together, through two different but coupled graphical means: a plot showing the variation of the current over time in a given cross-section of the wire and a 2D vector representation, as it varies in both time and space, of the electric field intensity vector around the wire induced (generated) by the current, with the two graphical representations being viewed simultaneously in a movie.

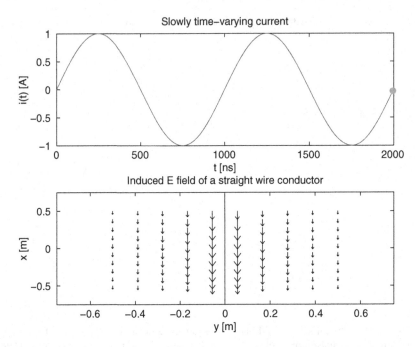

FIGURE 4.12 Another example from MATLAB assignment 2 illustrating the power of MATLAB computation and visualization in electromagnetics (fields): Snapshot of a movie in MATLAB that shows how the induced electric field intensity vector due to a time-harmonic current in a finite straight wire conductor varies in time and space, along (simultaneously) with the temporal variation of the current (two subplots are viewed simultaneously in the movie).

4.3 RESULTS AND DISCUSSION

Due to many changes associated with the RED project in the junior year of the EE program at CSU in the Fall of 2016, we decided that there would be simply too much work for the students if they had additional MATLAB homework project assignments every week or every other week on top of everything else required of them due to these changes. As such, the electromagnetics (fields) instructor only gave two separate large multi-week MATLAB assignments that were optional for extra course credit. Consequently, a quantitative analysis of how these assignments affected the students' grades is not appropriate. Instead, at the end of each MATLAB assignment, the students were given a survey to complete, which measured their thoughts, opinions, reactions, and feelings about how the assignment affected their understanding and enjoyment of both electromagnetics (fields) and MATLAB itself.

In particular, the students who chose to complete the MATLAB assignments were asked to complete a survey regarding their perceptions on the use of MATLAB and its integration into the curriculum. The surveys were voluntary and anonymous and the students were told that we would look at the results of the surveys only in aggregate form. The survey questions shown below were created by the course instructor. The students responded using an unbalance odd Likert scale. The following values were used:

- Yes, a lot (3)
- Yes, Somewhat (2)
- Not At All (0)
- No Opinion (Not Counted)

The first survey had a response/participation rate of 46% (39 out of 85 students in class), while the second survey only had a response rate of 22% (19 out of 85 students). Table 4.3 shows the results of the surveys with the average response (AR) and coefficient of variation (CV) presented. The students were also given the opportunity to make any comments about the project without any prompting questions.

Many students commented on the lack of MATLAB preparation: "I thought it was a good assignment that had significance to electromagnetics and MATLAB use, but I was very confused since I have never used MATLAB before." "I really enjoyed doing the MATLAB assignment, but I also found it to be very challenging especially question number 1.15 that one was very hard." "The MATLAB assignment was a good idea, because MATLAB is a very useful program to use and learn. This assignment was very difficult, considering a lot of students are not very familiar with MATLAB and all of its delicacies. I liked the idea of MATLAB, but some of the problems were difficult to complete with my knowledge of the program." "I thought it was a fun project, it helped to learn new things in MATLAB that I didn't know you could do. It was very challenging though typing up the code when some of us are unfamiliar with MATLAB since we've only used it a hand full of times in other classes." "I understand why you would like us to know and use MATLAB in this

TABLE 4.3

Student Survey Response Results, with the Average Response (AR) and Coefficient of Variation (CV) Presented for Each of the Two Large Multi-Week MATLAB Electromagnetics (Fields) Assignments

Student Survey Questions	Assignment 1		Assignment 2	
	AR	CV	AR	CV
1. Did MATLAB Exercises in parallel with traditional written problems and conceptual questions help you develop a stronger intuition and a deeper understanding of electromagnetic field theory, examples, and problems?	1.65	0.66	2.11	0.43
2. Did MATLAB Exercises help you find Electromagnetic Fields more attractive and likable?	1.46	0.73	1.50	0.86
3. Did MATLAB Exercises help you improve your knowledge and skill of using MATLAB, that is, help you build and enhance the understanding and command of MATLAB syntax, functionality, and programming, in the framework and cause of electromagnetics?	2.08	0.44	1.89	0.60
4. Did MATLAB Exercises allow you to gain some operational knowledge and skills in concepts and techniques of MATLAB use and programming that could potentially be used and implemented in other courses in the curriculum?	1.98	0.44	2.18	0.44
5. Did you find this MATLAB Project assignment challenging?	2.60	0.29	2.42	0.32

course but the reality is that most students I know don't have a good if any foundation using MATLAB. It's a great tool and is used a lot outside of the classroom for multiple applications. I think students need to have a MATLAB course to help enhance our understanding of the program."

Even students with substantial previous MATLAB experience, through research projects or internships, found many challenges and learned a lot from the completion of these projects. For example, "I personally have a ton of experience working with MATLAB, both for robotic applications as well as an entire data visualization program for a company" but "I had never been introduced to meshes before and I'm sure that they are a valuable tool that I will use in the future. I learned about plotting vector meshes as well and some other related methods." Also, "Very useful to learn that such tools as surf and quiver3 exists it MATLAB. As well as surfnorm."

Many students acknowledged and appreciated the importance of MATLAB as an essential tool for ECE and the need for gaining or improving MATLAB use and programming familiarity and expertise for students and engineers: "I like what you are trying to do with this project because I do believe that it will be important for all of us to be familiar with MATLAB at some point before we start our careers. I think it was a good call making the assignment extra credit because most of us have very little understanding with MATLAB in the first place." "I like this assignment a lot and you should definitely offer this project in future. It already has help me in another class, ..." "I appreciate the idea of incorporating MATLAB into the curriculum, specifically since MATLAB is so widely used in scientific fields." "I thought it was a nice change from the normal homework. I believe MATLAB is very important for electrical engineers so I am glad we are using it." "I personally think that the MATLAB project was a great idea as in industry I used MATLAB all of the time."

Some students confirmed that MATLAB exercises helped them develop a stronger intuition and a deeper understanding of electromagnetic field theory, examples, and problems: "Overall I like that you are trying to enhance our knowledge of electromagnetics using MATLAB because I think it's a very important skill to have and it is interesting to see some of the figures and graphs created using MATLAB. It helps visualize how certain changes and other variables affect the output and fields." "I think that this assignment is a really neat way to combine my knowledge of computer programming to a subject that I am not fond of." "I feel that the MATLAB assignment was very useful for giving us a better visual understanding of how Electromagnetics works." "This project helps me a lot about how to use MATLAB to plot and solve questions systematically." "This homework was really useful to understand the problems a lot better. It will be useful to have one or two MATLAB problems with every homework...." "I think this project was useful but more so for familiarizing us with MATLAB than improving our knowledge of Emag. It was good to reinforce what we already know because I knew what to expect the answers to be."

On the other hand, many students stated that MATLAB exercises helped them (instead) gain or improve operational knowledge and skills in concepts and techniques of MATLAB use and programming: "How MATLAB is used in the case of this assignment or other ECE MATLAB assignments, has taught me more about MATLAB than the actual MATLAB class did." "I feel like the MATLAB assignment helped me learn, mainly in MATLAB. My understanding of Electromagnetics did not significantly improve, although the problems required knowledge of electromagnetics. However, my MATLAB knowledge improved greatly because my knowledge prior was very basic." "It enhances my MATLAB programming a lot."

More specifically, the first survey results show that the students felt that the assignment really helped them to improve their knowledge and skills in MATLAB. They also felt that the assignment might be useful in other courses in the curriculum. The students did not feel strongly about the exercises really helping them develop a stronger intuition and a deeper understanding of electromagnetic field theory. This is likely due to the large struggle in learning MATLAB simultaneously with doing the assignment.

After the students had gained the knowledge and skills in the first MATLAB assignment, the second survey shows that they felt less strongly that the assignment improved this area – that is, their knowledge and skills in MATLAB. However, the students did feel relatively strongly that the exercises did indeed help them gain a stronger intuition and deeper understanding of electromagnetic field theory. This supports the idea that the students likely did not gain the understanding benefit due to a lack of knowledge and skills in the required software and programming. This will be something to keep in mind for future iterations of this research. The students also gained an insight into how the skills they improved in programming MATLAB could also apply to other courses as well. This is once again likely to be because the students are no longer focusing on learning the skills of MATLAB, but can instead realize the broader implications of their activities.

In addition, the survey results need to be viewed and judged in the context of the entire RED program and the increased workload for the students in the individual core junior-level EE courses and their LSMs, in the associated KI modules between the courses, foundations and professionalism sessions, and team activities. Many students found the "extra" MATLAB assignments overwhelming and overly time-demanding in the context of all the other newly established and required course and program components. "I never had a chance to really try all the problems with multiple other time constraints." "I feel this assignment would have been better to assign when time wasn't so constrained. I barely had any time to work on it and even then I felt very rushed."

Finally, in addition to the above qualitative analysis of the new approach and project results through student feedback surveys, we also used the results on the Electromagnetics Concept Inventory (EMCI) [41,42]. Namely, the instrument was administered at the start of the Fall 2016 and the Fall 2017 semesters, respectively, to the seniors who studied electromagnetics in the previous academic year. The 2017 score, by students learning electromagnetic fields in Fall 2016, was 2.5 times higher than the score by students taught without MATLAB Creativity Thread and MATLAB programming assignments.

4.4 CONCLUSIONS

This paper has presented and discussed the inclusion of computer-assisted MATLAB-based instruction and learning in the electromagnetics (fields) course and LSMs within the creativity thread of the REvolutionizing engineering and computer science

Departments project in the ECE Department at Colorado State University. Creativity class sessions were followed by two comprehensive and rather challenging multi-week homework assignments of MATLAB problems and projects in electromagnetics. This is enabled by a unique and extremely comprehensive collection of MATLAB tutorials, exercises, projects, and codes, developed by one of the faculty team members. These materials exploit technological and pedagogical power of MATLAB software as a general learning technology and constitute a modern tool for learning electromagnetics via computer-mediated exploration and inquiry. The assignments required the students to solve problems and generate results presented in figures, diagrams, movies and animations, by writing, testing, and executing their own MATLAB programs as well as running existing codes. The goal for students as learners is to gain and solidify the knowledge of fundamentals of electromagnetic fields using MATLAB and to build and enhance the understanding and command of MATLAB syntax, functionality, and programming in the framework of electromagnetics.

The principal difference of the presented approach when compared to the previous studies is the focus on introduction of MATLAB programming of electromagnetics performed by students, whereas in the previous reports, the students primarily conducted virtual experiments interacting with a GUI based on a MATLAB platform rather than writing their own MATLAB programs. The difference is also in the scope of the engaged computer-assisted MATLAB-based instruction and learning of electromagnetics in terms of topics, concepts, and techniques, with our developed MATLAB tutorials, exercises, and projects covering and reinforcing practically all important theoretical concepts, methodologies, and problem-solving techniques in electromagnetics and most of the previous studies focusing on a much smaller subset of topics. In addition, none of the previous studies with a similar intent, have, to the best of our knowledge, reported the results of the study in terms of analyses of the impact of the proposed and implemented approaches on students' performance, learning, mastery, attitude, success, and satisfaction that would enable comparison with the study presented in this paper.

The paper has presented and discussed some example problems from the Creativity Thread MATLAB assignments related to some of the most important pedagogical features of MATLAB pertaining to electromagnetic fields instruction/learning, such as manipulation and visualization of vectors and multivariable spatial functions, numerical analysis, and symbolic programming. It has shown the benefits of solving the problems and studying the topics both analytically and using MATLAB. The paper has also discussed how MATLAB assignments enable engagement of students' learning styles less frequently reached by typical lecture type of teaching and standard simulation software and tools, including visual, active, global, and inductive learning styles, most often helping several different categories of student learners at the same time.

Due to multifaceted changes introduced in association with the RED project in the junior year of the EE program at CSU in the Fall of 2016, we concluded that a quantitative analysis of how these MATLAB assignments affected the students' grades was not appropriate or feasible. Instead, as done in other prior studies, the results of this project were qualitatively analyzed using data from surveys given to the students at the end of each MATLAB assignment. The students who participated in the MATLAB exercises and surveys had mixed opinions about whether it was helpful to their personal understanding and mastery of the material. Overall, however, the students had relatively positive feedback for the assignments. There was a rather large percentage (72% on the first assignment and 63% on the second) of the students who even took the time to provide comments in addition to answering the five questions on the survey. For example, of the students who commented, 19 of the 28 had positive things to say about the first assignment. The majority of the negative comments all stem from a perceived lack of previous experience in MATLAB that was nearly universal among the students. But this is exactly why these MATLAB sessions and assignments were included, through the electromagnetics course, in the creativity thread of the RED project, i.e., to improve students' operational knowledge and skills in concepts and techniques of MATLAB use and programming, which can then be utilized in other courses in the curriculum, as well as in research activities, senior design (capstone) projects, etc. In addition, students' results on the Electromagnetics Concept Inventory assessment instrument showed great improvement when compared to the approach with no MATLAB sessions and assignments of the previous year.

ACKNOWLEDGEMENT

This work was supported by the National Science Foundation under grant EEC-1519438. The authors would like to thank Andrea Leland for her editorial changes that have greatly improved the presentation of this work.

REFERENCES

1. T. Chen, A. A. Maciejewski, B. M. Notaros, A. Pezeshki, and M. D. Reese, *"Mastering the Core Competencies of Electrical Engineering through Knowledge Integration,"* *Proceedings of the 123rd American Society for Engineering Education Annual Conference & Exposition – ASEE2016*, New Orleans, LA, June 26–29, 2016.

2. A. A. Maciejewski, T. W. Chen, Z. S. Byrne, M. A. de Miranda, L. B. Sample McMeeking, B. M. Notaros, A. Pezeshki, S. Roy, A. M. Leland, M. D. Reese, A. H. Rosales, R. F. Toftness, and O. Notaros, "A Holistic Approach to Transforming Undergraduate Electrical Engineering Education," Special Section "Innovations in Electrical and Computer Engineering Education", *IEEE Access*, Vol. 5, pp. 8148–8161, 2017.

3. S. R. H. Hoole, "A Course on Computer Modeling for Second or Third Year Engineering Undergraduates," *IEEE Transactions on Education*, vol. 36, no. 1, pp. 79–89, 1993.

4. B. Beker, D.W. Bailey, and G. J. Cokkinides, "An Application-Enhanced Approach to Introductory Electromagnetics," *IEEE Transactions on Education*, vol. 41, no. 1, pp. 31–36, 1998.

5. M. Chetty, S. Hu, and J. Bennett, "An Interactive Java-Based Educational Module in Electromagnetics," *International Journal of Electrical Engineering Education*, vol. 40, no. 1, pp. 79–90, 2003.

6. F. T. Ulaby and B. L. Hauck, "Undergraduate Electromagnetics Laboratory: An Invaluable Part of the Learning Process," *Proceedings of the IEEE*, vol. 88, no. 1, pp. 55–62, 2000.

7. I. D. Dinov, J. Sanchez, and N. Christou, "Pedagogical Utilization and Assessment of the Statistic Online Computational Resource in Introductory Probability and Statistics Courses," *Journal of Computers and Education*, vol. 50, pp. 284–300, 2008

8. R. M. Felder and J. Spurlin, "Applications, Reliability and Validity of the Index of Learning Styles," *International Journal of Engineering Education*, vol. 21, no. 1, pp. 103–112. 2005.

9. R. M. Felder and L. K. Silverman, "Learning and Teaching Styles in Engineering Education," *Engineering Education*, vol. 78, no. 7, pp. 674–681, 1988.

10. L. E. Bernold, J. E. Spurlin, and C. M. Anson, "Understanding Our Students: A Longitudinal-Study of Success and Failure in Engineering with Implications for Increased Retention," *Journal of Engineering Education*, vol. 96, no. 3, pp. 263–274. 2007.

11. R. M. Felder and R. Brent, "Understanding Student Differences," *Journal of Engineering Education*, vol. 94, no. 1, pp. 57–72, 2005.

12. T. A. Litzinger, S. H. Lee, J. C. Wise, and R. M. Felder, "A Psychometric Study of the Index of Learning Styles©," *Journal of Engineering Education*, vol. 96, no. 4, pp. 309–319, 2007.

13. R. M. Felder, R. Brent, and M. J. Prince, "Engineering Instructional Development: Programs, Best Practices, and Recommendations," *Journal of Engineering Education*, vol. 100, no. 1, pp. 89–122, 2011.

14. E. Seymour and H. Hewitt, *Talking About Leaving: Why Undergraduates Leave the Sciences*. Boulder, CO: Westview Press, 1997.

15. R. J. Wlodkowski, *Enhancing Adult Motivation to Learn: A Comprehensive Guide for Teaching All Adults*. San Francisco, CA: John Wiley & Sons, 2011.

16. National Research Council, *How People Learn: Brain, Mind, Experience, and School*. Expanded edition. Washington, DC: National Academies Press, 2000.

17. A. A. Read, "Computers and Computer Graphics in the Teaching of Field Phenomena," *IEEE Transactions on Education*, vol. 33, no. 1, pp. 95–103, 1990.

18. M. F. Iskander, "NSF/IEEE CAEME center: An exciting opportunity to align electromagnetics education with the nineties," *Computer Applications in Engineering Education*, vol. 1, no. 1, pp. 33–44, 1992.

19. W. L. Stutzman, "Integration of the personal computer into undergraduate electromagnetic courses," *Computer Applications in Engineering Education*, vol. 1, no. 3, pp. 223–226, 1993.

20. J. C. Rautio, "Educational use of a microwave electromagnetic analysis of 3-D planar structures," *Computer Applications in Engineering Education*, vol. 1, no. 3, pp. 243–254, 1993.

21. M. F. Iskander, J. C. Catten, R. Jameson, A. Jones, and A. Balcells, "Development of multimedia modules for education," *Computer Applications in Engineering Education*, vol. 3, pp. 97–110, 1995.

22. M. F. Iskander, J. C. Catten, R. M. Jameson, A. Rodriguez-Balcells, and A. K. Jones, "Interactive multimedia CD-ROMs for education," *Computer Applications in Engineering Education*, vol. 4, pp. 51–60, 1996.

23. J. C. Rautio, "The impact on education of widely available commercial 3-D planar electromagnetic software," *Computer Applications in Engineering Education*, vol. 8, no. 2, pp. 51–60, 2000.

24. S. Sanz, M. F. Iskander, and L. Yu, "Development of an interactive multimedia module on antenna theory and design," *Computer Applications in Engineering Education*, vol. 8, pp. 11–17, 2000.

25. M. F. Iskander, "Technology-based electromagnetic education," *IEEE Transactions on Microwave Theory and Techniques*, vol. 50, no. 3, pp. 1015–1020, Mar 2002.
26. K. C. Gupta, T. Itoh and A. A. Oliner, "Microwave and RF education-past, present, and future," *IEEE Transactions on Microwave Theory and Techniques*, vol. 50, no. 3, pp. 1006–1014, Mar 2002.
27. M. de Magistris, "A MATLAB-Based Virtual Laboratory for Teaching Introductory Quasi-Stationary Electromagnetics," *IEEE Transactions on Education*, vol. 48, no. 1, pp. 81–88, 2005.
28. S. Selleri, "A MATLAB Experimental Framework For Electromagnetic Education," *IEEE Antennas and Propagation Magazine*, vol. 45, no. 5, pp. 86–90, 2003.
29. J. M. Bértolo, F. Obelleiro, J. M. Taboada, and J. L. Rodríguez, "General purpose software package for electromagnetics engineering education," *Computer Applications in Engineering Education*, vol. 10, pp. 33–44, 2002.
30. A. W. Glisson and A. Z. Elsherbeni, "An interactive 1D Matlab FDTD code for education," *Computer Applications in Engineering Education*, vol. 9, pp. 136–147, 2001.
31. C. Mias, "Electronic Problem Based Learning of Electromagnetics through Software Development," *Computer Applications in Engineering Education*, vol. 16, no. 1, pp. 12–20, 2008.
32. L. Sevgi, *"Teaching electromagnetics via virtual tools,"* 2015 *IEEE International Symposium on Antennas and Propagation & USNC/URSI National Radio Science Meeting*, Vancouver, BC, 2015, pp. 1021–1022.
33. O. Ozgun and L. Sevgi, "VectGUI: A MATLAB-Based Simulation Tool [Testing Ourselves]," *IEEE Antennas and Propagation Magazine*, vol. 57, no. 3, pp. 113–118, June 2015.
34. H. G. Espinosa and D. V. Thiel, "MATLAB-Based Interactive Tool for Teaching Electromagnetics [Education Corner]," *IEEE Antennas and Propagation Magazine*, vol. 59, no. 5, pp. 140–146, Oct. 2017.
35. V. Demir and S. Kist, *"Matlab demonstrations for concepts in electromagnetics,"* 31st *International Review of Progress in Applied Computational Electromagnetics (ACES)*, Williamsburg, VA, pp. 1–2, 2015.
36. G. Apaydin and L. Sevgi, "A MATLAB-Based Virtual Tool for Simulations of Wave Propagation Inside a Parallel-Plate Waveguide [Testing Ourselves]," *IEEE Antennas and Propagation Magazine*, vol. 59, no. 4, pp. 100–105, Aug. 2017.
37. B. M. Notaroš, *MATLAB®-Based Electromagnetics*. Upper Saddle River, NJ: Pearson Prentice Hall, 2013.
38. B. M. Notaroš, R. McCullough, S. B. Manić, and A. A. Maciejewski, *"Work in Progress: Introducing MATLAB-Based Instruction and Learning in the Creativity Thread of a Novel Integrated Approach to ECE Education,"* *Proceedings of the 124th American Society for Engineering Education Annual Conference & Exposition – ASEE2017*, Columbus, OH, June 25–28, 2017.
39. B. M. Notaroš, *Electromagnetics*. Upper Saddle River, NJ: Pearson Prentice Hall, 2010.
40. J. F. Hoburg, *"Can Computers Really Help Students Understand Electromagnetics?,"* *Joint MMM-Intermag Conference Proceedings*, Pittsburgh, PA, 1991.
41. B. M. Notaros, *"Concept Inventory Assessment Instruments for Electromagnetics Education,"* *Proceedings of 2002 IEEE Antennas and Propagation Society International Symposium*, Vol. 1, pp. 684–687, San Antonio, TX, June 16–21, 2002.
42. D. L. Evans, D. Gray, S. Krause, J. Martin, C. Midkiff, B. M. Notaros, M. Pavelich, D. Rancour, T. Reed-Rhoads, P. Steif, R. Streveler, and K. Wage, *"Progress on Concept Inventory Assessment Tools,"* *Proceedings of the 33rd ASEE/IEEE Frontiers in Education Conference - FIE 2003*, pp. T4G.1–8, November 5–8, 2003, Boulder, CO.

5 Interactive Computational Tools for Electromagnetics Education Enhancement

Hugo G. Espinosa

Griffith University, Australia

Levent Sevgi

Istanbul OKAN University, Turkey

CONTENTS

5.1 INTRODUCTION

The subject material in electromagnetics courses is very mathematical, highly conceptual, and demands significant dedication to understand the concepts. While textbooks used in these courses provide some support, most of the exercises consist of forward modeling that provides solutions to the equations, rather than exercises that require the student to solve the problem and provide the solution. Thus, students can explore the relevant equations via the outputs of the mathematical models without needing to engage in solving the equations themselves. A variety of approaches to enhance electromagnetics education have been developed in the past decades, aiming to increase student engagement and improve the quality of education by reinforcing the understanding of complex concepts. Some of these approaches include theoretical alternatives to the use of vector analysis [1]; novel and alternative course plans and programs [2–5]; simulation strategies for complex electromagnetic systems, both analytic and numerical [6]; and different computer-assisted educational material [7–11]. The latter approach includes MATLAB-based computational tools such as a virtual laboratory for teaching quasi-stationary fields [12], visualization of electromagnetic wave reflection and refraction [13], vector analysis and curvilinear coordinate system [14], fault detection and identification of transmission lines [15], and even touch-based interactive tools through mobile technology for the analysis of coupled-line structures using the multiple-1D coupled-line finite-difference time-domain method [16]. The columns "Testing ourselves" and "Education corner" from the *IEEE Antennas and Propagation Magazine* continuously showcase virtual tools for electromagnetic analysis and modeling [17–22]; some of these tools can be accessed on-line or have been made available for download. In addition to their use in classical lectures, such tools were also developed to support alternative teaching methods, such as problem-based learning. These computational resources share common objectives: to enhance the teaching practices of instructors and increase student skills and engagement, enhance student learning outcomes, and improve knowledge retention.

MATLAB-based tools can be used as part of blended learning teaching approaches and can be integrated with existing online learning initiatives. One such initiative is massive (> 100,000 students) open online courses (MOOCs), which are freely available on-line from prestigious universities [23]. These courses provide, among others, interactive user forums, recorded lectures, online textbooks, quizzes, and practice problems, evolving as a distance education system. Together with the use of initiatives such as MOOCs, the development of appropriate virtual computational tools allows universities to improve how electromagnetics courses might be taught.

Four computational tools are presented in this chapter. Details regarding their development, computer engine, graphical user interface (GUI), applications, and case studies are included. In some instances, the effectiveness of the tool through anonymous student feedback is included. These tools have been published in the literature, and the majority of them are available to the teaching and scientific community through different platforms, such as stand-alone versions available for download, or web-based virtual laboratories. These tools have been developed in the MATLAB® environment (www.mathworks.com). The aim of this chapter is to offer the reader (instructor and/or student) a range of virtual and interactive resources that can be used to complement their electromagnetics courses. An instructor can use these resources to enhance their teaching and to explain physical and mathematical concepts more interactively. A student can use these tools to increase their engagement and interest in electromagnetics education.

First, a MATLAB-based interactive tutorial tool using a randomized problem generator and solver for electromagnetics courses is presented [24]. Examples of electrostatics, magnetostatics, dynamic fields, and transmission lines problem sets are included. Results from assessment of student feedback on its effectiveness are discussed. Second, a MATLAB-based virtual time-domain reflectometer, for the detection and identification of faults along finite-length transmission lines is presented [15]. The third tool consists of a MATLAB-based antenna package for the visualization of radiation characteristics of planar arrays of isotropic radiators. The beamforming capabilities for different locations, the number of radiators, as well as the different operating frequencies can be visualized [25]. A ray-shooting MATLAB-based package for the visualization of ray paths for ground-wave propagation simulations is presented as the fourth tool. The package can model wave propagation through complex wave-guiding environments [26].

5.2 DESCRIPTION OF THE VIRTUAL TOOLS

5.2.1 MATLAB-BASED INTERACTIVE TOOL FOR TEACHING ELECTROMAGNETICS

A MATLAB-based interactive tutorial tool using a randomized problem generator and solver for electromagnetics courses has been developed. Students are required to log into the tool and attempt to solve numerical problems. Every problem is different, having randomized input data and question selection from an extensive list. In the event of an incorrect answer, a library of anticipated errors is used to provide feedback about the appropriate method to solve this type of problem. In the event of a correct answer, a report is automatically issued to the user and the information stored in a local database. It is possible to log student access and also to record the number of attempts made to solve problems. No additional marking is required, as the courseware provides the answers and, if thought appropriate, the relevant equations in the textbooks can be suggested.

The MATLAB-based interactive tool can also be used as a screening instrument embedded within courses. For instance, progression through the course could be made conditional on satisfactory completion of the problems on prerequisite knowledge. Entry to a laboratory experiment could be made conditional on completion of some prerequisite exercises. Also, international students can be evaluated in their country prior to their application of candidature being accepted.

5.2.1.1 Description of the Application

The interface was developed using an efficient GUI autogeneration MATLAB tool [27], which allows the creation of graphical interfaces without touching the GUI MATLAB code. A text file that defines all component properties, including the figures, its controls, and the callbacks, is created by single lines of text. The type of object is defined by a keyword followed by the object's name, position, and callback function or code to execute. Examples of an output using the autogeneration tool are shown in [28,29].

Upon running the application, the window shown in Figure 5.1 is presented.

The upper half of the window in Figure 5.1 contains the instructions on how to use the application. Students enter their details in the first frame and run the application. An individual validation code is assigned to them, and the second frame from the top of the window is enabled, which allows the selection of the relevant chapter and topic through a pop-up menu. The possible electromagnetics chapters include "Electrostatics," "Magnetostatics," "Dynamic Fields," and "Transmission Lines." Table 5.1 contains the selected topics for the corresponding chapter. The tool has been designed in a way that with simple modifications and adjustments of the problem engine, it can be adapted to a variety of engineering courses.

Once the chapter and topic are selected, a random problem is generated by displaying an additional window (Figure 5.2). The top frame of the window includes the description of the problem, the problem's geometry, and the respective question. The second frame allows the student to enter his or her calculated values as scalars or vectors. A series of buttons allows the student to verify the solution, visualize features related to the problem, such as fields, charges, potential lines, current vectors, and so forth, and select another random problem.

Interactive Tool for Teaching Electromagnetics

Application developed for tutorial problems in electromagnetic fields and propagating systems.

Instructions:

1) Enter your name, student number and run the application.
2) Select one of the three chapters (Electrostatics, Magnetostatics, Dynamic fields and Transmission lines).
3) Select a topic that corresponds to the chapter.
4) Click on "Generate problem". A separate window will display the problem.
   Solve the problem and use at least 4 significant digits on your solutions.
   You can generate as many random problems as possible from that topic.
5) Once your solution is correct, click on "Save result".
6) In the main window, click on "Generate report" to generate a pdf file with your correct solutions.
7) Submit the report to your course convenor.

Student details

Name: _____   Student ID: _____   Run application

Problem selection

Select chapter: Electrostatics   Select topic: Coulomb's Law   Generate problem

Application Log

Logging date: 22/11/2020   Validation code: Problems solved correctly 0   Generate report

Griffith UNIVERSITY
Queensland, Australia

This application has been developed by Griffith School of Engineering.

Version 2.4 - 2020
Authors: Hugo G. Espinosa & David V. Thiel

Close application

FIGURE 5.1 The main application window of the MATLAB-based interactive tool.

TABLE 5.1

Chapter and Topics for Problem Selection

Electrostatics	Magnetostatics	Dynamic fields	Transmission Lines
• Charge density	• Forces and torques	• Faraday's Law	• Lumped-element model
• Coulomb's Law	• Biot-Savart Law	• Ideal transformer	• Transmission-line equations
• Gauss' Law	• Gauss' Law for magnetism	• Electromagnetic generator	• Wave propagation
• Electric scalar potential	• Ampere's Law	• Displacement current	• Lossless
• Conductors	• Magnetic potential	• Charge-current continuity	• trans line
			• Wave impedance
• Dielectrics	• Properties of materials	• Free-charge dissipation	• Power flow on lossless line
• Capacitance	• Inductance	• Electromagnetic potentials	• Smith chart
• Potential energy	• Magnetic energy		• Impedance matching
• Image theory			

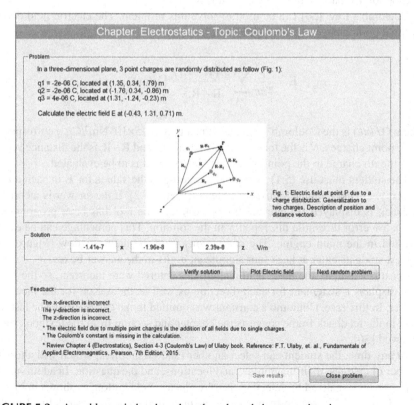

FIGURE 5.2 A problem window based on the selected chapter and topic.

The random problem feature provides an infinite selection of input parameters within the same problem, allowing the student to practice the same topic as many times as desired or required until they successfully complete the problem. If the solution is correct, the student saves the results. Once saved, the third frame in the main window (Figure 5.1) is enabled so the student can generate a PDF report for assessment. If the solution is incorrect, the bottom frame in Figure 5.2 provides feedback taking into account the most likely error, using a library of common errors encountered by students. Hints on how to solve the problem correctly are given as well as details on the topic/chapter from the reference book for review. The tool contains smart error detection, where the feedback is related to the specific error in the calculation, so guidance on how to correct the problem is provided. The student may re-attempt the problem as many times as they need in order to correctly solve the problem, although the input parameters will vary each time.

5.2.1.2 Example of an Electrostatics Problem

A sample problem is shown in Figure 5.2, where three-point charges are randomly distributed in free space. Their magnitudes are $-2\mu C$, $-2\mu C$, and $4\mu C$, and the student is required to calculate the vector of the electric field ($\mathbf{E}$) in a randomly selected point $\mathbf{P}$ in space. Note that the image in Figure 5.2 is a generalization of a 3-point charge distribution in space with random coordinates.

Coulomb's Law [30] can be used to solve this problem. The electric field due to multiple point charges can be determined by

$$\mathbf{E} = \frac{1}{4\pi\varepsilon} \sum_{i=1}^{N} \frac{q_i(\mathbf{R}-\mathbf{R}_i)}{|\mathbf{R}-\mathbf{R}_i|^3} \ (\text{V/m}) \tag{5.1}$$

where $(1/4\pi\varepsilon)$ is the Coulomb's constant equal to 8.9875×10^9 Nm^2/C^2, q corresponds to the point charges, N is the total number of charges, and $\mathbf{R}-\mathbf{R}_i$ is the distance vector from the ith charge to the point where the electric field is to be evaluated.

The student must use (5.1) to manually determine the values for $\mathbf{E}$ in each direction (x, y, z), and the field values are entered (Figure 5.2). If the answer is incorrect, feedback is displayed (Figure 5.2 bottom frame). The tool has been programmed with a 5% error threshold discrepancy in the solution. This percentage can be easily adjusted in the main engine. The instructions from the main window (Figure 5.1) suggest students enter at least four significant digits in the answer boxes.

For this example, the three field magnitudes entered were incorrect, so the feedback displays a statement for each direction as well as the reason for the incorrect answer. In this case, Coulomb's constant was omitted in the calculation. The last dot point in the feedback frame suggests the chapter and section from the textbook to be reviewed [30].

At any time, the student can select another random problem, which will alter the number of charges, their magnitudes and locations, and the question. In addition, the fields can be visualized by selecting the plot electric field button, as shown in Figure 5.3, where each charge corresponds to the magnitude and location defined in the

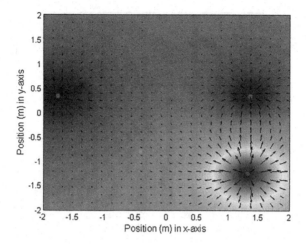

FIGURE 5.3 A visualization of the electric field vectors between the three-point charges from the problem depicted in Figure 5.2.

original problem from Figure 5.2. For a clearer display of the vector fields, only coordinates x and y are plotted.

If the correct **E** field values are entered, a pop-up message is displayed acknowledging the correct answer and enabling the option to generate the PDF report in the main window (Figure 5.1). In addition to the report form, the student's identification, validation code, chapter number, and topic number are stored in a local database for the record. The report can register as many entry problems as they were solved in one log in session. This form can be automatically submitted to the lecturer through platforms such as Blackboard learning management system[1]. More examples of electromagnetic problems are discussed in [24].

5.2.1.3 Course Overview

The interactive tool was developed to be used in the *Electromagnetic Fields and Propagating Systems* course for second-year undergraduate students in electronic and computer engineering at Griffith University, Nathan Campus, Queensland, Australia. The tool was provided to the students, so they could practice problems in their own time and at their own pace with the assistance of the textbook. The main textbook utilized in this course is *Fundamentals of Applied Electromagnetics*, by Fawwaz T. Ulaby and Umberto Ravaioli [30].

To evaluate the effectiveness of the tool, anonymous student feedback was collected through a survey form containing eight questions. For questions 1–7, students were given the option to select *yes, unsure,* or *no,* and they were also asked to provide reasons for their answers. The survey questions can be found in Appendix 5.1.

A total of 32 students participated in the survey (100% response rate). Figure 5.4 summarizes their responses; all students found the tool useful (Q1), and particularly valued the tool's capacity to provide immediate feedback on whether an answer they submitted was correct or incorrect (Q3). Students described the tool as an "alternative and fun way of learning" and a good tool for practicing problems for exams. The

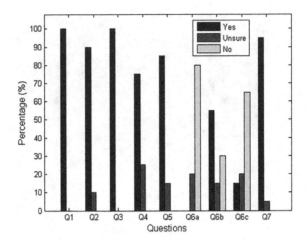

FIGURE 5.4 Anonymous survey responses from 32 students, questions 1–7.

majority found the hints (smart error detection) useful as they provided guidance on how to solve the problems correctly without revealing the answer straight away (Q2). The small proportion of students who responded with "unsure" to this question solved the problem correctly on their first attempt and, therefore, were not exposed to the feedback feature.

Students reported that having a visual representation of the fields, current vectors, potential lines, and so forth is a good way to understand the questions and a useful guidance to understand the concepts (Q4). As indicated in Figure 5.4, some students did not try this feature and responded with "unsure" to this question.

Furthermore, most of the students indicated that once they submitted a correct response they would continue solving additional random problems (Q5), enabling them to repeatedly practice the same topic with problems that are similar, albeit with varying questions and input parameters. They reported that for a different mistake, different feedback would be provided, which would enhance their understanding of the concept.

Regarding questions 6a, 6b, and 6c, although students found the tool useful, they believe it cannot replace lectures, tutorials, or laboratory questions. They reported that face-to-face lectures are the most important way of learning, as the lecturer will explain concepts in more detail and will address any question during the lecture time. However, several students believed that this tool could eventually replace tutorials, as it can be used anywhere and at any time.

The majority of the students would like to see this tool implemented in other courses (Q7). The tool has been developed as a series of structured MATLAB functions, so to adapt it to another course, only the functions that contain the engine (equations and calculations), together with the problem description, images, and feedback for every topic, would need to be adjusted. The GUI interface would remain the same.

Regarding Question 8, students reported that the computer-based tool offered many benefits. They indicated the tool was interactive and easy to use, allowed them to practice problems at their own pace, supported their learning by generating immediate feedback, provided graphical representations of the different concepts, contained a wide selection of topics with problems of varying levels of difficulty, and included a range of questions representing the whole problem space.

5.2.1.4 Summary

An interactive tool for assisting in teaching electromagnetics was implemented in MATLAB using a GUI autogeneration tool. This is an alternative to creating graphical interfaces without touching the GUI MATLAB code, by creating sets of text files. The tool includes four basic topics on electromagnetics: electrostatics, magnetostatics, dynamic fields, and transmission lines. The versatility and flexibility offered allows the tool to be adapted to other courses with easy adjustments on specific MATLAB functions. The tool offers unique features, such as random generation of an infinite number of problems, immediate feedback/hints when the answer is incorrect by using a smart error detection system, and report generation. This resource can be applied across the complete course and many other mathematically based courses with the following consequences: Greater engagement of students with the subject matter, improved student learning outcomes, improved prior knowledge assessment of potential students applying for university entry, and improved engagement for online students.

5.2.2 TRANSMISSION LINE FAULT ANALYSIS USING A MATLAB-BASED VIRTUAL TIME-DOMAIN REFLECTOMETER TOOL

Time-domain reflectometers (TDRs) are devices used to locate and identify faults along all types of cables such as broken conductors, water damage, cuts, smashed cables, short circuits, open circuits, and so forth. In the time-domain reflectometry principle, a generator injects a pulse down a transmission line (TL), and reflections from discontinuities and/or terminations are recorded. The distance between the generator and a fault is measured from the time delay between the incident pulse and the echo. Also, detailed analysis of the echo signal can reveal additional details of the faults or reflecting objects. A MATLAB-based virtual time-domain reflectometer (TDRMeter) tool was developed in [31] using the finite-difference time-domain (FDTD) derivation of the TL equations, by discretizing the source and load nodes under different termination conditions with the help of node-voltage and/or mesh-current circuit analysis techniques. The tool is applied here to the fault detection and identification along a TL.

5.2.2.1 Description of the Application and Examples

Upon running the TDRMeter application, the window shown in Figure 5.5 is presented.

There are four input data blocks on top of the panel. The user supplies the unit-length TL parameters based on its resistance (R_0), conductance (G_0), capacitance (C_0), and inductance (L_0) on the left block, from which the characteristic impedance of the line (Z_0) is automatically determined and displayed. The mid-left block is reserved for the generator parameters. Using the pop-up menu, the user can select

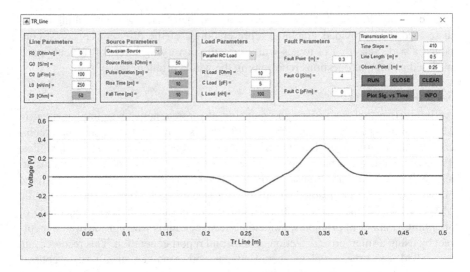

FIGURE 5.5 Front panel of the TDRMeter virtual simulation tool.

one of three different source types: Gaussian pulse, rectangular pulse, and a trapezoidal pulse and, if the trapezoidal source is selected, the pulse duration and rise/fall times need to be supplied. The internal source resistor is also supplied inside this block. The pulse length of the Gaussian voltage source is automatically selected according to user-specified line-length and other discretization parameters (as explained in [31]). The user may choose long pulse durations (i.e., at least longer than the total TL travel time) and simulate a step voltage source. The third block on the mid-right is used for the specification of the load. The pop-up menu of this block includes a resistive load, a parallel combination of resistor/capacitor, a serial combination of resistor/inductor, and serial and parallel resonance terminations. Based on the selection, the user is asked to supply resistor–inductor–capacitor (RLC) elements of the load in the activated data-boxes. The last block on the right-hand side is used for the fault specification. The user can change the unit-length capacitance and/or conductance. The TL length, observation point, and the number of simulation steps are supplied at the right top of the front panel, together with the runtime buttons.

The output of the TDRMeter is given via two different plots. The incident voltage pulse and reflected echoes (if they exist) along the TL as a function of length (in meters) are displayed on the bottom of the main window as movie frames (see Figure 5.5). Voltage plotted as a function of time in nanoseconds, at the specified observation point, may be displayed in a separate plot, which becomes visible when selecting the *Plot Sig. vs Time* button from the main window (see Figure 5.6). The same button automatically saves the data as *SigvsTime.dat* for further off-line signal analysis.

The TDRMeter can be used as either a *TL simulator*, or a *time-domain (TD) reflectometer*. In the TL simulator option, the user enters the load and fault parameters, while in the TD reflectometer option, the load and fault parameters are randomly selected, so that the user can determine those parameters by analysis and observation of the output signals.

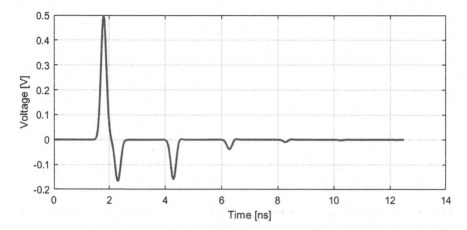

FIGURE 5.6 Voltage signal as a function of time in ns, excited by a Gaussian pulse generator. Six echoes observed due to a G-type fault located at 0.3 m from the generator.

The example shown in Figure 5.5 corresponds to a 50 Ω, 0.5 m loss-less TL, excited with a Gaussian pulse generator with 50 Ω internal resistance, and with unit length capacitance and inductance values given by 250 nH/m and 100 pF/m, respectively. The line is terminated with a parallel RC load, where the load resistance (R_L) and load capacitance (C_L) are given by 10 Ω and 5 pF, respectively. The number of steps is 500, and the observation point is located at 0.25 m of the line. A fault is introduced in the line at 0.3 m from the source, with a fault conductance of 4 S/m. The plot in Figure 5.5 clearly shows a negative reflection back to the generator at the 0.3 m discontinuity. The partially transmitted signal is reflected at the load back to the generator; part of the signal passes through the discontinuity and part of it reflects back to the load. This scenario can be observed in Figure 5.6, where six echoes can be easily identified.

The TDRMeter tool can be used in undergraduate electromagnetics courses to visualize the TD pulse characteristics and echoes from various discontinuities/terminations. Fault types and locations can be predicted. The simplest method is to predict the length of the TL (for the case of resistive discontinuities/termination) by measuring the transit time between the incident pulse and the echo (if separated in time). Distinguishing the end- and fault-reflected pulses in time, measuring delays among them, and marking the maximum amplitudes, are sufficient for this purpose.

An example is given in Figure 5.7. A 50 Ω, 0.5 m loss-less and fault-less TL with a load resistor of 150 Ω, is excited with a rectangular pulse, with 400 ps pulse duration. The distance from the observation point to the end of the line is determined by the time-delay between the two pulses. The distance from the generator to the observation point is determined by the transit time to the first pulse. The ratio of the pulse amplitudes is 0.25/0.5 = 0.5, which corresponds to the reflection coefficient (Γ). This value can easily be verified using the theoretical representation of Γ [30], such as

$$\Gamma = \frac{Z_L - Z_0}{Z_L + Z_0} = \frac{150 - 50}{150 + 50} = 0.5 \tag{5.2}$$

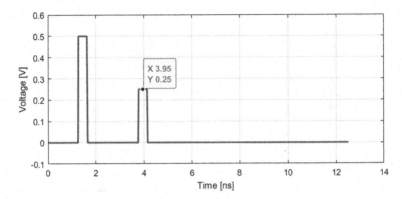

FIGURE 5.7 Voltage signal as a function of time in ns, excited with a rectangular pulse (pulse length 400 ps), with $Z_0 = 50 \, \Omega$ and $R_L = 150 \, \Omega$.

The previous example consisted of a purely real load; in order to solve problems with complex loads, the standard Fourier Transform is used by discriminating both the incident and reflected pulses in time, moving to the frequency domain by applying a fast Fourier transform (FFT), and calculating the ratio between the reflected pulse and the incident pulse within the source frequency band.

The shape of the echo may be used to predict the nature of the mismatch along the TL. One method is to use the Laplace transform to determine the time variation of the echo analytically. Another method is to observe the echo signal in time at the time limits. More theoretical details can be found in [15].

5.2.2.2 Plane Wave: Transmission Line Analogy

There is a close analogy between the propagation of plane waves in a homogeneous medium, and the propagation of voltage and current along a uniform transmission line. This analogy can be used not only as an aid to understand these concepts, but also to obtain solutions to realistic problems. A background in EM wave theory, for example, certainly helps to write down the solutions to EM wave problems, or vice versa.

For a uniform TL having primary parameters R, L, C, and G per unit length, the coupled TL equations for the voltage (v) and current (i) in differential forms can be re-arranged as [30]

$$\frac{\partial v(z,t)}{\partial z} + L\frac{\partial i(z,t)}{\partial t} + Ri(z,t) = 0 \tag{5.3a}$$

$$\frac{\partial i(z,t)}{\partial z} + C\frac{\partial v(z,t)}{\partial t} + Gv(z,t) = 0 \tag{5.3b}$$

while Maxwell equations for a uniform plane wave propagating along the +z direction inside a homogeneous medium, and having only E_x and H_y components can be written as [30]

$$\frac{\partial E_x(z,t)}{\partial z} + \mu \frac{\partial H_y(z,t)}{\partial t} = 0 \tag{5.4a}$$

$$\frac{\partial H_y(z,t)}{\partial z} + \varepsilon \frac{\partial E_x(z,t)}{\partial t} + \sigma E_x(z,t) = 0 \tag{5.4b}$$

where ε, μ, and σ correspond to the permittivity, permeability, and conductivity of the medium, respectively. Inspecting (5.3, 5.4) shows that the quantities in Table 5.2 are analogous.

TABLE 5.2

Analogous Parameters with Units between Circuit and Plane Wave Models

Circuit Model	Plane-wave Model
V [Volt]	E [Volt/m]
I [Amp]	H [Amp/m]
C [Farad/m]	ε [Farad/m]
L [Henry/m]	μ [Henry/m]
G [Siemens/m]	σ [Siemens/m]
R [Ohm/m]	$1/\sigma$

```
% ------------------------------------------------------------
% Program: LS_FDTD1D.m (Ex along Z-axis)
% ------------------------------------------------------------
clc; clear all; close all;
Exoldr=0;                    % 1-time step previous value at right boundary (k=ke-1)
Exoldl=0;                    % 1-time step previous value at left boundary (k=1)
nt=1500; zmax=100; ke=100; kc=30; ep0=1/(36*pi*1e9); mu0=4*pi/1e7; c=3e8;
lbc=input('Left Boundary Termination: (1) PEC, (0) Free-space = ');
rbc=input('Right Boundary Termination: (1) PEC, (0) Free-space = ');

tt=ke/2; t0=10; delz=floor(zmax/ke); dt=delz/c;
band=c/(10*delz); alfa=3.3*band*band; shift=4./sqrt(alfa);
for k=1:ke     % put zeros into arrays
    Ex(k)=0.0; Hy(k)=0.0; z(k)=k;
end
t=0;
for n=1:nt     % FDTD time loop starts
    t=n*dt;
        for k=2:ke-1     % Ex is calculated alon z
            Ex(k)=Ex(k)+dt/(ep0*delz)*(Hy(k-1)-Hy(k));
        end
    Ex(kc)=Ex(kc)+exp(-alfa*(t-shift)^2);
    if lbc==0; Ex(1)=Exoldl;  end
    if rbc==0; Ex(ke)=Exoldr; end
    for k=1:ke-1     % Hy is calculated along z
        Hy(k)=Hy(k)+dt/(mu0*delz)*(Ex(k)-Ex(k+1));
    end
    Exoldr=Ex(ke-1); Exoldl=Ex(2); % Keep Ex for next time step
    plot(z, Ex); xlim([1,ke]); ylim([-1.1,1.1]); grid;
    xlabel('Z-axis [m]'); ylabel('Ex [V]'); pause(0.0001)
end % ------------------ END ------------------
```

FIGURE 5.8 a) 1D FDTD-based MATLAB script for plane wave simulations;
(Continued)

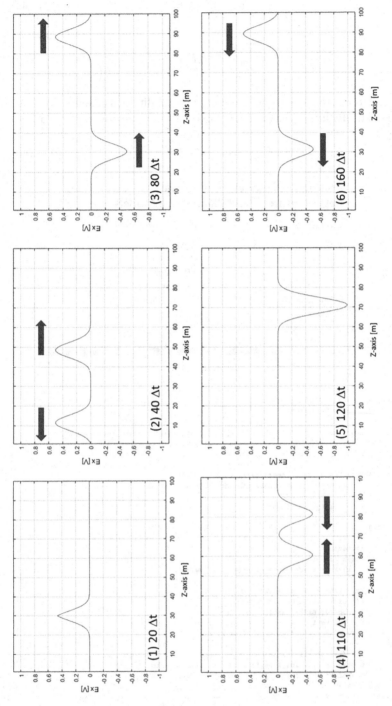

FIGURE 5.8 (Continued) b) snapshots of a plane wave propagating along the ±*Z*-direction in free space at different time instances.

The following MATLAB script (Figure 5.8a) allows the user to create movie clips for an FDTD-based one-dimensional time-domain plane-wave model. Figure 5.8b shows several snapshots from a movie clip of the evolution of a plane wave propagating along the ±Z-direction (in meters) in free space at different time instances. In this case, both the left and right boundary terminations of the model are defined as perfect electrical conductors (i.e., full reflections with opposite phase at both ends).

5.2.2.3 Summary

The TDRMeter virtual tool can be used to teach, understand, test, and visualize TD TL characteristics, and echoes from various discontinuities and terminations. Inversely, the types and locations of faults in the TL can be predicted from the recorded data. Reflections due to various types of transmission structures, impedance effects of vias, signal and power line couplings, etc., on printed circuits with small dimensions make signal integrity one of the most important and complex electromagnetic compatibility (EMC) problems. The virtual tool may be particularly helpful in the assessment of signal integrity. The flexibility of the tool allows the user to define and/or add other types of faults by simply modifying the MATLAB m-file. This simulation tool is freely available to the community upon request. In addition, the analogy between plane wave theory and transmission line theory has been presented.

5.2.3   A MATLAB-BASED VISUALIZATION PACKAGE FOR PLANAR ARRAYS OF ISOTROPIC RADIATORS

A MATLAB-based antenna array package (ANTEN_GUI) has been developed for the visualization of radiation patterns, beamforming, and beam-steering from planar arrays of isotropic radiators. The package has been designed to be used as an educational resource in undergraduate electromagnetics courses to enhance students' understanding of antenna concepts.

An antenna array refers to the arrangement of multiple radiators interconnected to produce a directional radiation pattern. The direction of the radiation pattern can be changed; this is referred to as beamforming or beam-steering. The steering can be achieved by either mechanical or electronic scanning. A group of antennas based on electronic scanning is called phase array. The GUI simulation tool presented here allows the visualization of two-dimensional and three-dimensional radiation patterns of several selected and user-located isotropic radiators (i.e., arrays).

5.2.3.1   Description of the Application and Examples

Upon running the ANTEN_GUI application, the window shown in Figure 5.9 is presented.

On the top left corner, the user can select the following options from the pop-up menu: *arbitrary, linear, planar,* or *circular arrays*, with a number of user-selected and located isotropic radiators. Once the number of radiators (N) and the maximum radius (in meters) of the polar region where the radiators will be located are specified, the user can select the button *locate radiators* and manually enter the location of each radiator. The engine uses the MATLAB built-in function *ginput* for the manual and arbitrary location of radiators. Once they have been located, the user can visualize the radiation pattern in either two- (button *plot*) or three-dimensions (button *3D plot*).

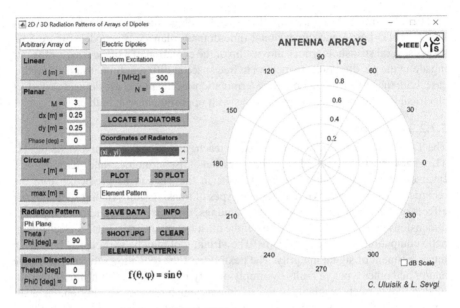

FIGURE 5.9   Simulation tool ANTEN_GUI main panel.

The radiation pattern plane is specified by the Theta (θ)/Phi (φ) plane pop-up menu and input box. If either the θ or φ planes are selected, the angle varies between $0°$ and $360°$, while the other angle remains fixed at a value specified by the user. The two-dimensional patterns are plotted with an angular resolution of $\Delta\theta = \Delta\varphi = 1°$.

The following examples (Figure 5.10) show the radiation pattern of four scenarios with different array configurations, radiation pattern planes, and beam directions.

As can be seen in Figure 5.10, the radiation patterns are determined by the geometrical configuration of the radiators, the radiation plane, and the beam direction. The user can choose from four different options: electric dipoles, magnetic dipoles, standing wave dipoles, and traveling wave dipoles, as well as between uniform and binomial excitation. Figure 5.10a shows the pattern of one radiator in the φ plane ($\theta = 90°$) with beam angles $\theta = 0°$ and $\varphi = 0°$. Figure 5.10b shows the pattern of a linear array of two radiators in the φ plane ($\theta = 90°$) with beam angles $\theta = 0°$ and $\varphi = 0°$. A linear array of seven radiators in the φ plane ($\theta = 90°$) with beam angles $\theta = 0°$ and $\varphi = 0°$ is shown in Figure 5.10c. Finally, a planar array of fourteen radiators in the φ plane ($\theta = 90°$) with beam angles $\theta = 90°$ and $\varphi = 0°$ is shown in Figure 5.10d. The user can design any kind of array when the option *arbitrary array* is selected; however, the software locates the radiators automatically when the options *linear, planar,* or *circular arrays* are selected. If the array type is linear, N elements are symmetrically located on the y-axis, with an inter-element distance defined by d. If the array type is planar, $N \times M$ elements are located on the x–y plane, where N corresponds to the number of elements along the y-direction, and M are the elements along the x-direction. The separation distances between the radiators are defined by dx and dy. If the array type is circular, N elements are symmetrically located on a circle with the center at the origin and a radius r defined by the user. The linear, planar, and circular input parameters are located on the left-hand side menus of Figure 5.9.

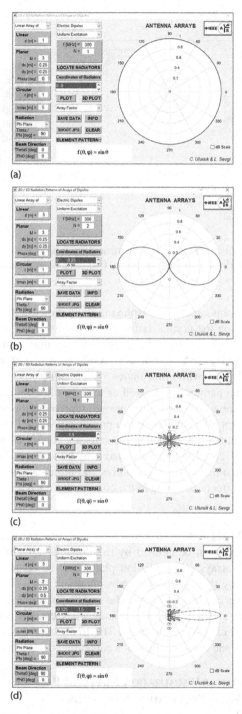

(a)

(b)

(c)

(d)

FIGURE 5.10 Radiation patterns of different array configurations, different radiation patterns, and beam direction characteristics: a) one radiator, b) linear array of two radiators, c) linear array of seven radiators, and d) planar array of fourteen radiators.

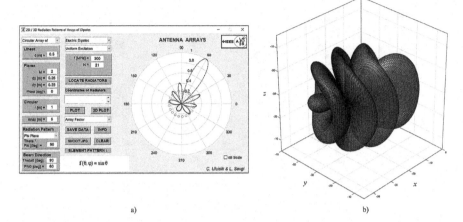

a)   b)

FIGURE 5.11   Radiation patterns of a 21-element circular array with $r =1$ m, in the φ plane ($\theta = 90°$) with beam angles $\theta = 90°$ and $\varphi = 60°$: a) two-dimensional radiation pattern, b) three-dimensional radiation pattern.

The simulation tool is capable of displaying three-dimensional radiation patterns. The engine is based on the normalization of the radius of the unit sphere for each observation angle (θ_i, φ_i), in a three-dimensional rectangular coordinate system. This feature can be seen in Figure 5.11, where a circular array of 21 radiators (radius $r =1$ m) in the φ plane ($\theta = 90°$) with beam angles $\theta = 90°$ and $\varphi = 60°$ is simulated. Figure 5.11a shows the input parameters, as well as the two-dimensional radiation pattern, while Figure 5.11b shows the three-dimensional representation of the radiation pattern. The frequency for all examples shown previously was set to 300 MHz.

More details about the theory behind the software engine, as well as more examples, can be found in [25]. The tool presented here, however, is a modified enhanced version of [25], which includes more features and more capabilities. This modified version of the tool is freely available to the community upon request.

5.2.3.2   Summary

An antenna-array MATLAB-based simulation tool was presented. Different geometrical configurations can be defined by the user; these include arbitrary, linear, planar, and circular arrays. The radiation plane and the beam direction are input parameters defined by the user. Two-dimensional and three-dimensional radiation pattern characteristics can be visualized. The simulation tool can be used as an educational resource in undergraduate electromagnetics courses, to enhance the student's understanding of antenna concepts. The freely available source code allows users to add features and improve the tool's capabilities.

5.2.4   A Ray-shooting Visualization MATLAB Package for 2D Ground-Wave Propagation Simulations

A MATLAB-based simulation tool (SNELL_GUI) for the visualization of ray paths through two-dimensional complex environments composed of obstacles and variable refractivity profiles is presented. The tool can be used to model wave propagation through complex wave-guiding environments, and as a design tool for the visualization of geometrical optics (GO) ray propagation, reflection, and refraction. The prediction of ray paths between the transmitter and receiver is a challenging task, so the objective of the tool is to determine a computationally fast way to specify the dominant ray paths that account for the field-strength prediction.

The SNELL_GUI tool shoots a number of rays; the angles of departure of such rays are specified by the user through a propagation medium characterized by several linear vertical refractivity profiles. The range-independent refractivity is assumed here, although its extension to range-dependent situations can be easily implemented. The core of the software engine is based on the direct application of Snell's Law between adjacent horizontal layers; the refractive indexes of such layers are assumed constant [30]. The ray angle is measured from the vertical axis.

The tool has been designed to draw ray paths through a complex environment, with a variable refractivity profile along a path that may or may not include obstacles. Up to two obstacles of different heights can be included in the simulations. In addition, the tool has been designed only for visualization purposes and for forward-propagating rays. The freely available source code, however, offers the flexibility to include electromagnetic field contributions of rays at a specified observation range and height, by using the theory described in [26]; that is, ray fields as a function of height with a specific range, or ray fields as a function of range with a constant height. The rays, emanating from the source, are shot one-by-one; the tool stores the points on the ray paths as (x, y) pairs, as long as the ray coordinates are lower than the maximum permitted range and height. In addition, the open-source code allows for the addition of backward-propagating ray visualization; this is particularly important when multiple forward and backward reflections (resonances) occur.

The tool has been programmed considering four different procedures:

1. Consecutive application of Snell's Law through a multi-horizontally-layered propagation medium. The height of each layer is constant and specified by the user in the input parameters. More details can be found in [26].
2. Full reflection from the perfect electrical conductor (PEC) bottom surface. In this case, the ray is bent upward with the same angle in which it hits the bottom surface.
3. Same as (2), but the reflection occurs on top of the obstacles (if specified by the user).
4. Reserved for instances in which ray hits the left wall of an obstacle; if this occurs, the ray ends there and the source shoots the next ray.

5.2.4.1   Description of the Application and Examples

Upon running the SNELL_GUI application, the window shown in Figure 5.12 is presented.

The left-hand side axis is used for the vertical refractivity profile ($n = n(x)$). The right-hand side axis (in yellow) is used to plot the rays. Source, refractivity, ray angle, and obstacle parameters are user-defined and entered in the input boxes at the top of the window.

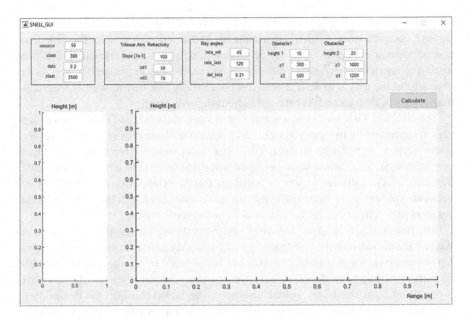

FIGURE 5.12 Front panel of the ray-shooting virtual simulation tool SNELL_GUI.

Figure 5.13 shows ray-shooting through an environment with a homogeneous atmosphere (i.e., $n = n_0 = 1$), with rays emanating from the source in the form of straight lines. The maximum height was defined as 300 m, with a source height of 50 m, and a layer height of 0.2 m. The maximum range was defined as 2500 m. The first and last ray angles were 45° and 120°, respectively, with a ray increment of 0.5°. Figure 5.13a shows an environment with no obstacles, while Figure 5.13b shows an environment with two obstacles. In this case, the height of one obstacle is 20 m with a base length of 200 m, and located at 500 m range, while the second obstacle is 85 m in height with a base length of 200 m, and located at 1500 m range. Reflections from the top of both obstacles (i.e., buildings) and rays ending at the left sides of the obstacles can be clearly observed in Figure 5.13b.

Figure 5.14 shows ray-shooting through an atmosphere with a typical trilinear refractivity profile; the slope of the refractivity was set to 0.001. The heights of ducting and anti-ducting were set to 80 m and 150 m, respectively. The maximum height, source height, maximum range, and layer height were set to the same parameters defined in Figure 5.13. The first and last ray angles were 45° and 120°, respectively, with a ray increment of 0.375°. Figure 5.14a shows an environment with no obstacles, while Figure 5.14b shows an environment with two obstacles. The parameters used to define the obstacles were set to the same parameters as in Figure 5.13b.

The effect of the obstacles (i.e., buildings) and the reflections occurring on top of them can clearly be seen in Figure 5.14b. It must be noted that not all the rays were plotted, only a finite number of rays in the range between 45° and 120°. Regions with multiple rays (i.e., the dark regions) may be either the field maximum or the field minimum, depending on the constructive or destructive ray interference.

An example of ray-shooting is presented in Figure 5.15. In order to reduce the effect produced by the second obstacle (85 m), with respect to the source height (50 m)

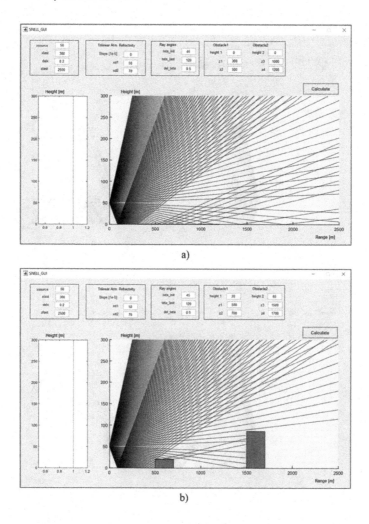

FIGURE 5.13 Ray-shooting visualization through an environment with homogeneous atmosphere: a) with no obstacles and b) with two obstacles. The top of each window shows the user-input parameters.

as shown in Figure 5.14, the source was extended to a 100 m height. All other parameters remained the same as the parameters defined in Figure 5.14.

More details about the theory behind the software engine, as well as more examples can be found in [26]. This version of the tool is freely available to the community upon request.

5.2.4.2 Summary

A ray-shooting simulation tool for the visualization of ray paths in a two-dimensional propagation environment, along simple surface-terrain profiles, with variable atmospheric refractivity was presented. The tool can be used to predict ducting and/or antiducting conditions in the presence of buildings and different atmospheric-refractivity

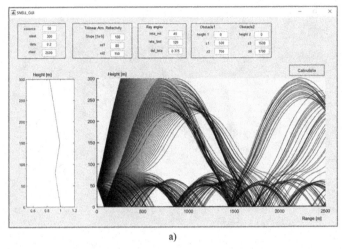

a)

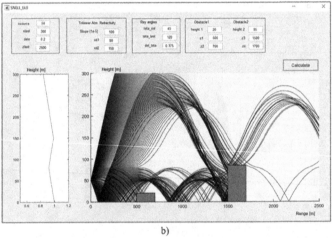

b)

FIGURE 5.14 Ray-shooting visualization through an atmosphere with a trilinear refractivity profile. Slope of refractivity = 0.001, heights of ducting and anti-ducting are 80 m and 150 m, respectively: a) environment with no obstacles and b) environment with two obstacles.

conditions. Undergraduate engineering courses focused on EM theory, antennas and propagation, and so forth, can benefit from using this resource. The freely available source code allows users to add features and improve the tool's capabilities.

5.3 CONCLUSIONS

Virtual interactive tools are effective resources that enhance students' understanding of EM concepts, increase their engagement, and improve their learning outcomes. Such tools are particularly important in the current climate, in which educators are experiencing an unprecedented demand for online teaching activities. Four computational MATLAB-based tools have been presented in this chapter, with several others

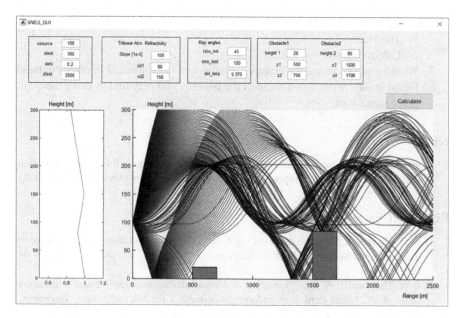

FIGURE 5.15   Ray-shooting visualization through an atmosphere with a trilinear refractivity profile. Slope of refractivity = 0.001, heights of ducting and anti-ducting are 80 m and 150 m, respectively. Two obstacles and a source height of 100 m.

recommended in the reference list. All tools have two objectives in common: to enhance the teaching practices of instructors, and to support the student's learning experience. One of the tools was based on the randomized selection of tutorial problems for practicing electromagnetics topics related to electrostatics, magnetostatics, dynamic fields, and transmission lines. The other three tools were based on simulation and visualization characteristics of topics related to time-domain reflectometry, antenna arrays, and propagation in complex environments. The tools are freely available to the community upon request.

APPENDIX 5.1

Survey questions for the tool: MATLAB-based interactive tool for teaching electromagnetics.

1. Did you find the automated tutorial useful?
2. Did you find the hints/feedback useful?
3. Is it important to get immediate feedback if the answer is correct or incorrect?
4. Did you find visualization graphics useful?
5. After a correct answer, would you continue using the automated tutorial to solve additional random problems?
6. Do you think this type of automated tutorial exercise could be used to replace any of the following?

  a. Lectures
  b. Tutorials
  c. Lab questions
 7. Would you like to see the computer-based tool implemented in other courses?
 8. Please briefly write down some of the benefits you may find in using the computer-based tool.

NOTE

1 Blackboard learning management system is a software application for the administration, documentation, tracking and reporting of educational technology (www.blackboard.com).

REFERENCES

1. K. F. Warnick, R. H. Selfridge, and D. V. Arnold, "Teaching electromagnetic field theory using differential forms," *IEEE Transactions on Education*, vol. 40, no. 1, pp. 53–68, Feb. 1997.
2. D. V. Thiel and J. W. Lu, "Numerical techniques in electromagnetics and communications. A PC based third year undergraduate subject for Microelectronic Engineers," *Proceedings of IEEE Antenna Propagation of Symposium*, pp. 111–114, 1994.
3. L. Sevgi, "A new electromagnetic engineering program and teaching via virtual tools," *Prog. in Electromagnetics Research B*, vol. 6, pp. 205–224, 2008.
4. L. Sevgi, "Teaching EM modeling and simulation as a graduate-level course," *IEEE Antennas and Propagation Magazine*, vol. 54, no. 5, pp. 261–269, 2012.
5. S. Y. Lim, "Education for electromagnetics: Introducing electromagnetics as an appetizer course for computer science and IT undergraduates," *IEEE Antennas and Propagation Magazine*, vol. 56, no. 5, pp. 216–222, 2014.
6. L. B. Felsen and L. Sevgi, "Electromagnetic engineering in the 21st century: challenges and perspectives," *ELEKTRIK, Turkish Journal of Electrical Engineering and Computer Sciences*, vol. 10, no. 2, pp. 131–145, 2002.
7. J. F. Hoburg, "Can computers really help students understand electromagnetics?," *IEEE Transactions on Education*, vol. 36, no. 1, pp. 119–122, 1993.
8. M. F. Iskander, "Technology-based electromagnetic education," *IEEE Trans. on Micro. Theory and Tech.*, vol. 50, no. 3, pp. 1015–1020, 2002.
9. S. C. Mukhopadhyay, "Teaching electromagnetics at the undergraduate level: a comprehensive approach," *Eur. J. Phys.*, vol. 27, pp. 727–742, 2006.
10. L. Sevgi, "Virtual tools/labs in electrical engineering education," *ELEKTRIK, Turkish Journal of Electrical Engineering and Computer Sciences*, vol. 14, no. 1, pp. 113–127, 2006.
11. L. Sevgi, "Electromagnetic modeling and simulation: challenges in validation, verification and calibration," *IEEE Transactions on EMC*, vol. 56, no 4, pp. 750–758, 2014.
12. M. de Magistris, "A MATLAB-based virtual laboratory for teaching introductory quasi-stationary electromagnetics," *IEEE Transactions on Education*, vol. 48, no. 1, pp. 81–88, 2005.
13. L. C. Trintinalia, "Simulation tool for the visualization of EM wave reflection and refraction," *IEEE Antennas and Propagation Magazine*, vol. 55, no. 1, pp. 203–211, 2013.
14. O. Ozgun and L. Sevgi, "VectGUI: A MATLAB-based simulation tool," *IEEE Antennas and Propagation Magazine*, vol. 57, no. 3, pp. 113–118, 2015.

15. L. Sevgi, "Transmission line fault analysis using a MATLAB-based virtual time domain reflectometer tool," *IEEE EMC Newsletter*, pp. 67–72, 2006.

16. E. L. Tan and D. Y. Heh, "Mobile teaching and learning of coupled-line structures," *IEEE Antennas and Propagation Magazine*, vol. 62, no. 2, pp. 62–69, 2020.

17. L. Sevgi and Ç. Uluışık, "A MATLAB-based Visualization Package for Planar Arrays of Isotropic Radiators," *IEEE Antennas and Propagation Magazine*, vol. 47, no. 1, pp. 156–163, Feb. 2005.

18. G. Çakır, M. Çakır, and L. Sevgi, "A Multipurpose FDTD-Based Two Dimensional Electromagnetic Virtual Tool," *IEEE Antennas and Propagation Magazine*, vol. 48, no. 4, pp. 142–151, Aug. 2006.

19. F. Akleman and L. Sevgi, "A Novel MoM- and SSPE-based Groundwave Propagation Field Strength Prediction Simulator," *IEEE Antennas and Propagation Magazine*, vol. 49, no. 5, pp. 69–82, Oct. 2007.

20. M. A. Uslu and L. Sevgi, "MATLAB-based Filter Design Program: From Lumped Elements to Microstriplines," *IEEE Antennas and Propagation Magazine*, vol. 53, no. 1, pp. 213–224, Feb. 2011.

21. A. Sefer, M. A. Uslu, and L. Sevgi, "MATLAB-based 3-D MoM and FDTD codes for the RCS analysis of realistic objects," *IEEE Antennas and Propagation Magazine*, vol. 57, no. 4, pp. 122–128, Aug. 2015.

22. Ç. Uluışık and L. Sevgi, "Web-based Virtual Laboratories for Antenna Arrays, Radiowave Propagation and Filter Design," *IEEE Antennas and Propagation Magazine*, vol. 53, no. 4, pp. 252–260, Aug. 2011.

23. Massachusetts Institute of Technology (2014). MIT OpenCourseWare [Online]. Available: http://ocw.mit.edu/index.htm. [accessed Nov. 25 2020].

24. H. G. Espinosa and D. V. Thiel, "MATLAB-based interactive tool for teaching electromagnetics," *IEEE Antennas and Propagation Magazine*, vol. 59, no. 5, pp. 140–146, 2017.

25. L. Sevgi and C. Uluisik, "A MATLAB-based visualization package for planar arrays of isotropic radiators," *IEEE Antennas and Propagation Magazine*, vol. 47, no. 1, pp. 156–163, 2005.

26. L. Sevgi, "A ray-shooting visualization MATLAB package for 2D ground-wave propagation simulations," *IEEE Antennas and Propagation Magazine*, vol. 46, no. 4, pp. 140–145, 2004.

27. H. G. Espinosa, D. A. James, S. Kelly, and A. Wixted, "Sports monitoring data and video interface using a GUI auto generation MATLAB tool," *Procedia Engineering (Elsevier)*, vol. 60, pp. 243–248, 2013.

28. H. G. Espinosa, D. A. James, and A. Wixted, "Video digitising interface for monitoring upper arm and forearm rotation of cricket bowlers," *Proceedings of ASTN*, vol. 1, no. 1, 2013.

29. P. Jithu, H. G. Espinosa, and D. A. James, "An interactive tool for conditioning inertial sensor data for sports applications," *Journal of Fitness Research*, vol. 5, pp. 37–39, 2016.

30. F. T. Ulaby and U. Ravaioli, *Fundamentals of applied electromagnetics*, Essex, England: Pearson, 7th Ed., 2015.

31. L. Sevgi and C. Uluisik, "A MATLAB-based transmission-line virtual tool: finite-difference time-domain reflectometer," *IEEE Antennas and Propagation Magazine*, vol. 48, no. 1, pp. 141–145, 2006.

6 Computational Electromagnetics and Mobile Apps for Electromagnetics Education

Eng Leong Tan

Nanyang Technological University, Singapore

CONTENTS

6.1 INTRODUCTION

Electromagnetics (EM) forms an important branch of physics that is indispensable to electrical and electronic engineers. Many applications are directly or indirectly related to EM theories, ranging from antenna design, electromagnetic compatibility/interference (EMC/EMI), wireless communications, microwave circuit design, and radio frequency (RF) engineering to laser and optoelectronics, etc. Therefore, teaching EM courses in tertiary level electrical engineering becomes paramount in order to help students understand the underlying physics and develop physical insight for problem-solving. However, it remains challenging to date for teaching EM courses in the tertiary level due to its abstract nature. Many EM theories also require high mathematical finesse and involve vectorial analyses in three-dimensional (3-D) space, which are often difficult to visualize and grasp by many students. Over the years, many initiatives and efforts have been undertaken to improve the teaching and learning of EM. Most approaches include the utilization of computer software, electronic books, and virtual experiments to elucidate various EM concepts [1–11]. These approaches provide certain extent of visualizations and interactions mostly on computers or over the web.

To enhance teaching and learning of electromagnetics, time-domain simulations provide much intuitional insight for wave propagations and dynamics. As such, time-domain computational electromagnetics (CEM) methods are more useful to illustrate various wave phenomena in physically intuitive manner compared to other methods in wavelet or polynomial-based domains. One of the most popular CEM methods in time domain is finite-difference time-domain (FDTD) method [12,13]. However, the conventional FDTD method is an explicit scheme with the time step size restricted by Courant–Friedrichs–Lewy (CFL) stability constraint. To overcome the CFL constraint, unconditionally stable implicit FDTD methods have been developed, such as alternating direction implicit (ADI), split-step (SS), locally one-dimensional (LOD) FDTD methods, etc. [14–18]. These unconditionally stable FDTD methods comprise more complicated implicit update procedures that involve matrix operators at both left- and right-hand sides (RHS) of update equations. To improve the computational efficiency, fundamental schemes for implicit FDTD methods have been developed that involve concise, simple, and efficient update equations with matrix-operator-free RHS [19–22]. With their unconditional stability and enhanced efficiency, time-domain simulations are "fast-forwardable" using time step size larger than CFL constraint. Such fast-forwardable feature is very useful for quick concept demonstrations and illustrations during teaching and learning.

Thus far, most of the computer- or web-based software and simulations still lack touch-based interactivity and real-time responsiveness. They are also deemed insufficient due to the limited visualization through two-dimensional (2-D) computer monitor screen. Furthermore, the requirement and over reliance of computer or web connection greatly inhibits seamless teaching and learning anytime, anywhere. In recent years, the wide availability of mobile devices such as smartphones or iPads has provided great opportunity for their utilization to enhance EM education. We have developed several EM educational apps on mobile devices to aid teaching and learning of various topics on EM wave propagation, wave interactions, transmission lines, etc. [23–29]. In order to enable EM simulations in real time on mobile devices, there is a need for innovative CEM methods that are efficient and require only small computing resources. However, since most 3-D full-wave CEM methods above call for huge computing resources and long simulation time, they are difficult to be implemented directly on the mobile device platform. To alleviate the difficulty, we have developed the multiple one-dimensional (M1-D) FDTD and coupled-line (CL) FDTD methods for multiple and coupled transmission lines [28–32]. The methods reduce the complexity of governing equations into multiple 1-D ones that are simpler and more concise. Hence, the M1-D methods are more manageable in terms of computing resources on mobile devices. They can run efficiently to provide interactive simulations and insightful visualizations in real time. They allow quick initial design and analysis of microwave circuits on mobile devices anytime, anywhere [33].

This chapter presents the CEM methods and mobile apps for EM education. In Section 6.2, some CEM methods in time domain are reviewed including the conditionally stable explicit 3-D FDTD method and unconditionally stable implicit 3-D FDTD methods. For the latter, the fundamental ADI and LOD FDTD methods are highlighted for their concise, simple, and efficient update equations with matrix-operator-free RHS. Further developments using more efficient explicit and implicit M1-D FDTD and CL-FDTD methods are also discussed. In Section 6.3, several educational mobile apps are described for enhanced teaching and learning of EM on various topics including EM polarization, plane wave reflection and transmission, microstrip circuits, transmission lines, and coupled-line structures. These mobile apps feature convenient and effective touch-based interactivity to aid teaching and learning. They are incorporated with innovative CEM methods that are efficient and fast-forwardable using unconditionally stable implicit methods. Interactive simulations and insightful visualizations of wave propagation are provided on the mobile devices to elucidate the concepts of EM theory, transmission lines, and microwave circuits. Besides mobile devices, 3-D displays may be supplemented for enhanced EM education as well. In particular, 3-D TV can provide the necessary depth perception for enhancing visualization of electromagnetic waves in 3-D space. Some educational study and survey results using mobile devices/3-D TV are discussed.

6.2 COMPUTATIONAL ELECTROMAGNETICS FOR EM EDUCATION

In this section, we review some CEM methods in time domain including the explicit and implicit 3-D FDTD methods. For applications in EM education, we also discuss the developments using more efficient explicit and implicit M1-D FDTD and CL-FDTD methods.

6.2.1 Explicit FDTD Method

Let us express the 3-D EM equations in compact matrix form as

$$\frac{\partial \mathbf{u}}{\partial t} = \mathbf{W}\mathbf{u} \qquad (6.1)$$

where $\mathbf{u}$ is the vector of electromagnetic fields and $\mathbf{W}$ is the 6×6 system matrix:

$$\mathbf{u} = \begin{bmatrix} \mathbf{E} & \mathbf{H} \end{bmatrix}^T = \begin{bmatrix} E_x & E_y & E_z & H_x & H_y & H_z \end{bmatrix}^T \qquad (6.2)$$

$$\mathbf{W} = \begin{bmatrix} 0 & \mathbf{W}_{12} \\ \mathbf{W}_{21} & 0 \end{bmatrix} = \begin{bmatrix} 0 & 0 & 0 & 0 & \dfrac{-1}{\epsilon}\dfrac{\partial}{\partial z} & \dfrac{1}{\epsilon}\dfrac{\partial}{\partial y} \\[2mm] 0 & 0 & 0 & \dfrac{1}{\epsilon}\dfrac{\partial}{\partial z} & 0 & \dfrac{-1}{\epsilon}\dfrac{\partial}{\partial x} \\[2mm] 0 & 0 & 0 & \dfrac{-1}{\epsilon}\dfrac{\partial}{\partial y} & \dfrac{1}{\epsilon}\dfrac{\partial}{\partial x} & 0 \\[2mm] 0 & \dfrac{1}{\mu}\dfrac{\partial}{\partial z} & \dfrac{-1}{\mu}\dfrac{\partial}{\partial y} & 0 & 0 & 0 \\[2mm] \dfrac{-1}{\mu}\dfrac{\partial}{\partial z} & 0 & \dfrac{1}{\mu}\dfrac{\partial}{\partial x} & 0 & 0 & 0 \\[2mm] \dfrac{1}{\mu}\dfrac{\partial}{\partial y} & \dfrac{-1}{\mu}\dfrac{\partial}{\partial x} & 0 & 0 & 0 & 0 \end{bmatrix} \qquad (6.3)$$

Equations (6.1–6.3) can be solved using the popular FDTD method with leapfrog time-stepping [12,13]:

$$\begin{bmatrix} \mathbf{E}^{n+\frac{1}{2}} \\ \mathbf{H}^{n+1} \end{bmatrix} = \left(\mathbf{I} + \Delta t \begin{bmatrix} 0 & 0 \\ \mathbf{W}_{21} & 0 \end{bmatrix} \right) \left(\mathbf{I} + \Delta t \begin{bmatrix} 0 & \mathbf{W}_{12} \\ 0 & 0 \end{bmatrix} \right) \begin{bmatrix} \mathbf{E}^{n-\frac{1}{2}} \\ \mathbf{H}^n \end{bmatrix} \qquad (6.4)$$

where Δt is the time step size and $\mathbf{I}$ is the identity matrix. The spatial derivatives can be discretized via central differencing on staggered Yee grids. Since the FDTD method is an explicit scheme, the time step size is restricted by the CFL stability constraint, i.e.,

$$\Delta t \le \Delta t_{\text{CFL}} = \frac{\Delta \sqrt{\epsilon \mu}}{\sqrt{3}} \qquad (6.5)$$

where Δ is the minimum mesh size. This constraint may severely limit the computational efficiency when there exist fine meshes in the simulation domain.

6.2.2 IMPLICIT FDTD METHODS

6.2.2.1 Fundamental ADI (FADI) FDTD Method

To overcome the stability constraint, unconditionally stable ADI-FDTD method has been introduced with two update procedures as [14,15].

$$\left(I - \frac{\Delta t}{2} A\right) u^{n+\frac{1}{2}} = \left(I + \frac{\Delta t}{2} B\right) u^n \tag{6.6}$$

$$\left(I - \frac{\Delta t}{2} B\right) u^{n+1} = \left(I + \frac{\Delta t}{2} A\right) u^{n+\frac{1}{2}} \tag{6.7}$$

where

$$A = \begin{bmatrix} 0 & 0 & 0 & 0 & 0 & \frac{1}{\epsilon}\frac{\partial}{\partial y} \\ 0 & 0 & 0 & \frac{1}{\epsilon}\frac{\partial}{\partial z} & 0 & 0 \\ 0 & 0 & 0 & 0 & \frac{1}{\epsilon}\frac{\partial}{\partial x} & 0 \\ 0 & \frac{1}{\mu}\frac{\partial}{\partial z} & 0 & 0 & 0 & 0 \\ 0 & 0 & \frac{1}{\mu}\frac{\partial}{\partial x} & 0 & 0 & 0 \\ \frac{1}{\mu}\frac{\partial}{\partial y} & 0 & 0 & 0 & 0 & 0 \end{bmatrix} \tag{6.8}$$

$$B = \begin{bmatrix} 0 & 0 & 0 & 0 & \frac{-1}{\epsilon}\frac{\partial}{\partial z} & 0 \\ 0 & 0 & 0 & 0 & 0 & \frac{-1}{\epsilon}\frac{\partial}{\partial x} \\ 0 & 0 & 0 & \frac{-1}{\epsilon}\frac{\partial}{\partial y} & 0 & 0 \\ 0 & 0 & \frac{-1}{\mu}\frac{\partial}{\partial y} & 0 & 0 & 0 \\ \frac{-1}{\mu}\frac{\partial}{\partial z} & 0 & 0 & 0 & 0 & 0 \\ 0 & \frac{-1}{\mu}\frac{\partial}{\partial x} & 0 & 0 & 0 & 0 \end{bmatrix} \tag{6.9}$$

Note that the RHS of (6.6–6.7) comprise matrix operators **A** and **B** that call for considerable floating-point operations (flops) count.

To improve the computational efficiency, we have developed the fundamental ADI (FADI) FDTD method with the following update procedures [19,20]:

$$\mathbf{v}^n = \mathbf{u}^n - \mathbf{v}^{n-\frac{1}{2}} \tag{6.10}$$

$$\left(\frac{1}{2}\mathbf{I} - \frac{\Delta t}{4}\mathbf{A}\right)\mathbf{u}^{n+\frac{1}{2}} = \mathbf{v}^n \tag{6.11}$$

$$\mathbf{v}^{n+\frac{1}{2}} = \mathbf{u}^{n+\frac{1}{2}} - \mathbf{v}^n \tag{6.12}$$

$$\left(\frac{1}{2}\mathbf{I} - \frac{\Delta t}{4}\mathbf{B}\right)\mathbf{u}^{n+1} = \mathbf{v}^{n+\frac{1}{2}} \tag{6.13}$$

where $\mathbf{v}$'s are the auxiliary field vectors that are only temporary and do not incur extra memory arrays. Unlike (6.6–6.7), the RHS of (6.10–6.13) contain only vectors and are matrix-operator-free (no more $\mathbf{A}$ or $\mathbf{B}$). The advantages of FADI-FDTD method include concise update equations that lead to simple, convenient coding and efficient implementation of implicit scheme. With matrix-operator-free RHS, they will also enable concise, simple, and efficient implementation of current sources [21]. Due to the unconditional stability of FADI-FDTD, one may use time step size larger than CFL constraint to achieve higher computational efficiency over explicit FDTD.

Many classical implicit finite-difference schemes comprise update procedures with various RHS matrix operators and even their sum and/or product. These classical implicit schemes include Douglas scheme, D'Yakonov scheme, Douglas–Gunn scheme, Crank–Nicolson scheme, etc. [33,34]. Applying the same principle of FADI to omit their RHS matrix operators, they can be transformed to the same form involving update procedures with matrix-operator-free RHS. In particular, equations (6.10–6.13) provide their simplifications into concise and efficient forms with matrix-operator-free RHS (no more $\mathbf{A}$ or $\mathbf{B}$) [19]. They constitute the basis of unification for many classical implicit schemes, providing insights into their inter-relations along with simplifications, concise updates, and efficient implementations.

6.2.2.2   Fundamental LOD (FLOD) FDTD Method

When high accuracy is not needed, such as during teaching and learning demonstration or initial design and analysis, alternative unconditionally stable methods have been developed that may be more efficient or more stable. One such method is the split-step (SS) or locally one-dimensional (LOD) FDTD method [16–18]:

$$\left(\mathbf{I} - \frac{\Delta t}{2}\mathbf{A}\right)\mathbf{u}^{n+\frac{1}{2}} = \left(\mathbf{I} + \frac{\Delta t}{2}\mathbf{A}\right)\mathbf{u}_1^n \tag{6.14}$$

$$\left(\mathbf{I}-\frac{\Delta t}{2}\mathbf{B}\right)\mathbf{u}_1^{n+1}=\left(\mathbf{I}+\frac{\Delta t}{2}\mathbf{B}\right)\mathbf{u}^{n+\frac{1}{2}} \tag{6.15}$$

This method is accurate to first order in time, hence the main field vectors are subscripted with "1" as $\mathbf{u}_1$. The RHS of (6.14–6.15) also comprise matrix operators $\mathbf{A}$ and $\mathbf{B}$ that involve considerable flops.

To improve the computational efficiency, the fundamental LOD (FLOD) FDTD method has been developed as [19,22,35]

$$\left(\frac{1}{2}\mathbf{I}-\frac{\Delta t}{4}\mathbf{A}\right)\mathbf{v}^{n+\frac{1}{2}}=\mathbf{u}_1^n \tag{6.16}$$

$$\mathbf{u}^{n+\frac{1}{2}}=\mathbf{v}^{n+\frac{1}{2}}-\mathbf{u}_1^n \tag{6.17}$$

$$\left(\frac{1}{2}\mathbf{I}-\frac{\Delta t}{4}\mathbf{B}\right)\mathbf{v}^{n+1}=\mathbf{u}^{n+\frac{1}{2}} \tag{6.18}$$

$$\mathbf{u}_1^{n+1}=\mathbf{v}^{n+1}-\mathbf{u}^{n+\frac{1}{2}} \tag{6.19}$$

Note that the RHS of the update procedures have been simplified in concise and efficient matrix-operator-free forms (no more $\mathbf{A}$ or $\mathbf{B}$). To achieve second-order temporal accuracy, careful treatments via input and output processings can be implemented as and when necessary at the desired observation locations [18], [36–38].

Further developments of the above ADI and SS/LOD FDTD methods can be carried out more conveniently using their fundamental implicit schemes with matrix-operator-free RHS [39–50]. Note that the LOD-FDTD method remains stable even for non-uniform (varying) time-steps [51]. Other implicit methods including ADI-FDTD tend to become unstable unless the time-step is uniform during run-time [52,53]. Besides two split matrices for the SS/LOD FDTD method, it is possible to have three split matrices for the system matrix in (6.1), i.e., $\mathbf{W}=\mathbf{A}_3+\mathbf{B}_3+\mathbf{C}_3$. The FLOD-FDTD methods with three split matrices have also been developed, which comprise four implicit update procedures with matrix-operator-free RHS [54,55].

6.2.3   M1-D Explicit FDTD Methods

6.2.3.1   M1-D FDTD Method for Transmission Lines and Stubs

The above CEM methods involve full-wave 3-D computations that have been targeted mainly for high-end computers or workstations. These "heavy-weight" CEM methods typically call for high-performance CPU and huge memory. They are not suitable for "light-weight" portable devices that are of limited resources with lower CPU capability and less memory. In order to enable real-time EM simulations, such as for interactive teaching and learning anytime, anywhere, there is a need for innovative CEM methods that are not only efficient, but also well-suited for implementations on mobile devices.

To bypass the computationally intensive full-wave 3-D methods, we have developed M1-D FDTD method for transmission lines, stubs and lumped elements. The method consists of simple M1-D update equations, which can be readily implemented on the platform of mobile devices. In particular, for main transmission lines (TLs) and stubs, their update equations are given as [28,29]

Main TLs:

$$E_x^m\Big|^{n+1} = E_x^m\Big|^n - \frac{\Delta t}{\epsilon}\frac{\partial}{\partial z}H_y^m\Big|^{n+\frac{1}{2}} - \frac{\Delta t}{\epsilon}J_x^m\Big|^{n+\frac{1}{2}} \tag{6.20}$$

$$H_y^m\Big|^{n+\frac{3}{2}} = H_y^m\Big|^{n+\frac{1}{2}} - \frac{\Delta t}{\mu}\frac{\partial}{\partial z}E_x^m\Big|^{n+1} \tag{6.21}$$

Stubs:

$$E_x^s\Big|^{n+1} = E_x^s\Big|^n + \frac{\Delta t}{\epsilon}\frac{\partial}{\partial y}H_z^s\Big|^{n+\frac{1}{2}} \tag{6.22}$$

$$H_z^s\Big|^{n+\frac{3}{2}} = H_z^s\Big|^{n+\frac{1}{2}} + \frac{\Delta t}{\mu}\frac{\partial}{\partial y}E_x^s\Big|^{n+1} \tag{6.23}$$

where

$$\epsilon = \frac{1}{Z_0 v}, \quad \mu = \frac{Z_0}{v}, \quad v = \frac{c}{\sqrt{\epsilon_{\text{eff}}}} \tag{6.24}$$

$$J_x^m\Big|^{n+\frac{1}{2}} = -\frac{\partial}{\partial y}H_z^s\Big|^{n+\frac{1}{2}} \quad \text{at interjunctions.} \tag{6.25}$$

E_x^m and H_y^m are the electric and magnetic fields of the mth main transmission line, while E_x^s and H_z^s are the electric and magnetic fields of the sth stub. Z_0, v and ϵ_{eff} are the transmission line characteristic impedance, phase velocity and effective permittivity, respectively.

The M1-D explicit FDTD method is useful for initial design, quick analysis and enhanced learning of transmission line circuits. However, it is conditionally stable with time step size restricted by stability constraint

$$\Delta t \le \frac{\Delta}{v\sqrt{2}} \tag{6.26}$$

where Δ is the minimum mesh size. This constraint may limit the computational efficiency and necessitate long simulations of wave propagation transients at times.

6.2.3.2 M1-D CL-FDTD Method for Coupled Transmission Lines

For coupled transmission lines, the differential equations are different from the ordinary 3-D, 2-D, or 1-D Maxwell's equations:

$$\frac{\partial \mathbf{u}_{cl}}{\partial t} = \mathbf{W}_{cl}\mathbf{u}_{cl}, \quad \mathbf{u}_{cl} = \left[E_{x1}, E_{x2}, H_{y1}, H_{y2}\right]^T \tag{6.27}$$

$$\mathbf{W}_{cl} = \begin{bmatrix} 0 & 0 & \dfrac{-1}{\epsilon_s}\dfrac{\partial}{\partial z} & \dfrac{-1}{\epsilon_m}\dfrac{\partial}{\partial z} \\[2mm] 0 & 0 & \dfrac{-1}{\epsilon_m}\dfrac{\partial}{\partial z} & \dfrac{-1}{\epsilon_s}\dfrac{\partial}{\partial z} \\[2mm] \dfrac{-1}{\mu_s}\dfrac{\partial}{\partial z} & \dfrac{1}{\mu_m}\dfrac{\partial}{\partial z} & 0 & 0 \\[2mm] \dfrac{1}{\mu_m}\dfrac{\partial}{\partial z} & \dfrac{-1}{\mu_s}\dfrac{\partial}{\partial z} & 0 & 0 \end{bmatrix}. \tag{6.28}$$

$\mathbf{W}_{cl}$ is the 4 × 4 coupled-line system matrix, E_{x1} and H_{y1} are the EM fields along line 1, while E_{x2} and H_{y2} are those along line 2. Equations (6.27–6.28) are also different from those used for uncoupled transmission lines in [28–31].

In conjunction with (6.27–6.28), we have developed the M1-D CL-FDTD method with the following update equations [56,57]:

$$E_{x1}^{n+1} = E_{x1}^{n} - \frac{\Delta t}{\epsilon_s}\frac{\partial}{\partial z}H_{y1}^{n+\frac{1}{2}} - \frac{\Delta t}{\epsilon_m}\frac{\partial}{\partial z}H_{y2}^{n+\frac{1}{2}} \tag{6.29}$$

$$E_{x2}^{n+1} = E_{x2}^{n} - \frac{\Delta t}{\epsilon_s}\frac{\partial}{\partial z}H_{y2}^{n+\frac{1}{2}} - \frac{\Delta t}{\epsilon_m}\frac{\partial}{\partial z}H_{y1}^{n+\frac{1}{2}} \tag{6.30}$$

$$H_{y1}^{n+\frac{3}{2}} = H_{y1}^{n+\frac{1}{2}} - \frac{\Delta t}{\mu_s}\frac{\partial}{\partial z}E_{x1}^{n+1} + \frac{\Delta t}{\mu_m}\frac{\partial}{\partial z}E_{x2}^{n+1} \tag{6.31}$$

$$H_{y2}^{n+\frac{3}{2}} = H_{y2}^{n+\frac{1}{2}} - \frac{\Delta t}{\mu_s}\frac{\partial}{\partial z}E_{x2}^{n+1} + \frac{\Delta t}{\mu_m}\frac{\partial}{\partial z}E_{x1}^{n+1} \tag{6.32}$$

where

$$\frac{1}{\epsilon_s} = \frac{v_e Z_{0e} + v_o Z_{0o}}{2}, \quad \frac{1}{\epsilon_m} = \frac{v_e Z_{0e} - v_o Z_{0o}}{2} \tag{6.33}$$

$$\frac{1}{\mu_s} = \frac{1}{2}\left(\frac{v_o}{Z_{0o}} + \frac{v_e}{Z_{0e}}\right), \quad \frac{1}{\mu_m} = \frac{1}{2}\left(\frac{v_o}{Z_{0o}} - \frac{v_e}{Z_{0e}}\right) \tag{6.34}$$

$$v_e = \frac{c}{\sqrt{\epsilon_{eff}^e}}, \quad v_o = \frac{c}{\sqrt{\epsilon_{eff}^o}} \tag{6.35}$$

ϵ_s, ϵ_m, μ_s and μ_m are the self and mutual permittivities and permeabilities, which can be expressed in terms of even- and odd-mode characteristic impedances (Z_{0e}, Z_{0o}), phase velocities (v_e, v_o), and effective permittivities (ϵ_{eff}^e, ϵ_{eff}^o) [58–60].

The M1-D CL-FDTD method requires only spatial discretization along the z direction, i.e., Δz. It bypasses the fine meshing of line width and spacing of the coupled transmission lines in the full wave 3-D FDTD method that are more computationally intensive. However, since it is still an explicit method, there is a stability constraint that restricts the time step size as

$$\Delta t \leq \frac{\Delta z}{\sqrt{\left|\dfrac{1}{\epsilon_s \mu_s} - \dfrac{1}{\epsilon_m \mu_m}\right| + \left|\dfrac{1}{\epsilon_s \mu_m} - \dfrac{1}{\epsilon_m \mu_s}\right|}} \tag{6.36}$$

Such time step limit for coupled transmission lines in (6.36) appears to be more restrictive than that for the single transmission line with $\Delta t \leq \Delta z \sqrt{\epsilon \mu}$.

6.2.4   M1-D Implicit FDTD Methods

6.2.4.1   M1-D FADI-FDTD Method for Transmission Lines and Stubs

To overcome the stability constraint, we have developed the unconditionally stable M1-D FADI-FDTD method. The method calls for one-step update procedures for main transmission lines and stubs as [61]

Main TLs:

$$\left(\frac{1}{2}\mathbf{I} - \frac{\Delta t}{4}\mathbf{B}_m\right)\mathbf{v}_m^{n+1} = \mathbf{u}_m^n \tag{6.37}$$

$$\mathbf{u}_m^{n+1} = \mathbf{v}_m^{n+1} - \mathbf{u}_m^n \tag{6.38}$$

Stubs:

$$\left(\frac{1}{2}\mathbf{I} - \frac{\Delta t}{4}\mathbf{A}_s\right)\mathbf{v}_s^{n+\frac{1}{2}} = \mathbf{u}_s^{n-\frac{1}{2}} \tag{6.39}$$

$$\mathbf{u}_s^{n+\frac{1}{2}} = \mathbf{v}_s^{n+\frac{1}{2}} - \mathbf{u}_s^{n-\frac{1}{2}} \tag{6.40}$$

where

$$\mathbf{B}_m = \begin{bmatrix} 0 & \dfrac{-1}{\epsilon_m}\dfrac{\partial}{\partial z} \\ \dfrac{-1}{\mu_m}\dfrac{\partial}{\partial z} & 0 \end{bmatrix}, \quad \mathbf{u}_m^{n+1} = \begin{bmatrix} E_x^m\Big|^{n+1} \\ H_y^m\Big|^{n+1} \end{bmatrix} \tag{6.41}$$

$$\mathbf{A}_s = \begin{bmatrix} 0 & \dfrac{1}{\epsilon_s}\dfrac{\partial}{\partial y} \\ \dfrac{1}{\mu_s}\dfrac{\partial}{\partial z} & 0 \end{bmatrix}, \quad \mathbf{u}_s^{n+\frac{1}{2}} = \begin{bmatrix} E_x^s\Big|^{n+\frac{1}{2}} \\ H_z^s\Big|^{n+\frac{1}{2}} \end{bmatrix} \tag{6.42}$$

The RHS of (6.37–6.40) for main TLs and stubs are matrix-operator-free, which simplify the implementations and improve the efficiency of implicit update procedures.

Using the M1-D FADI-FDTD method, the EM fields in all interconnected main TLs and stubs can be updated cooperatively and efficiently to solve practical circuits. The simulations can be accelerated by adjusting the time step size specified in terms of CFLN = $\Delta t/\Delta t_{CFL}$. Thanks to its unconditional stability, the simulations are "fast-forwardable" using time step size larger than CFL constraint, i.e., CFLN > 1. Alternative to FADI, one may also resort to FLOD-FDTD method with non-uniform time steps for better trade-off between efficiency and accuracy [62]. Both M1-D FADI and FLOD FDTD methods can be extended for transmission line circuits including lumped elements and networks [63–66].

6.2.4.2   M1-D FADI CL-FDTD Method for Coupled Transmission Lines

The coupled-line system matrix $\mathbf{W}_{cl}$ can be written as the sum of proper split matrices [67]:

$$\mathbf{W}_{cl} = \mathbf{A}_{cl} + \mathbf{B}_{cl} \tag{6.43}$$

$$\mathbf{A}_{cl} = \begin{bmatrix} 0 & 0 & \dfrac{-1}{\epsilon_s}\dfrac{\partial}{\partial z} & 0 \\ 0 & 0 & \dfrac{-1}{\epsilon_m}\dfrac{\partial}{\partial z} & 0 \\ \dfrac{-1}{\mu_s}\dfrac{\partial}{\partial z} & 0 & 0 & 0 \\ \dfrac{1}{\mu_m}\dfrac{\partial}{\partial z} & 0 & 0 & 0 \end{bmatrix} \tag{6.44}$$

$$\mathbf{B}_{\mathrm{cl}} = \begin{bmatrix} 0 & 0 & 0 & \dfrac{-1}{\epsilon_m}\dfrac{\partial}{\partial z} \\[2mm] 0 & 0 & 0 & \dfrac{-1}{\epsilon_s}\dfrac{\partial}{\partial z} \\[2mm] 0 & \dfrac{1}{\mu_m}\dfrac{\partial}{\partial z} & 0 & 0 \\[2mm] 0 & \dfrac{-1}{\mu_s}\dfrac{\partial}{\partial z} & 0 & 0 \end{bmatrix} \tag{6.45}$$

These split matrices will lead to stable schemes even for time step size larger than CFL constraint. Other sets of split matrices have been proposed and investigated but some of them may not preserve stability [68–70].

The M1-D FADI CL-FDTD method can be formulated for coupled transmission lines using the split matrices as

$$\mathbf{v}_{\mathrm{cl}}^n = \mathbf{u}_{\mathrm{cl}}^n - \mathbf{v}_{\mathrm{cl}}^{n-\frac{1}{2}} \tag{6.46}$$

$$\left(\frac{1}{2}\mathbf{I} - \frac{\Delta t}{4}\mathbf{A}_{\mathrm{cl}}\right)\mathbf{u}_{\mathrm{cl}}^{n+\frac{1}{2}} = \mathbf{v}_{\mathrm{cl}}^n \tag{6.47}$$

$$\mathbf{v}_{\mathrm{cl}}^{n+\frac{1}{2}} = \mathbf{u}_{\mathrm{cl}}^{n+\frac{1}{2}} - \mathbf{v}_{\mathrm{cl}}^n \tag{6.48}$$

$$\left(\frac{1}{2}\mathbf{I} - \frac{\Delta t}{4}\mathbf{B}_{\mathrm{cl}}\right)\mathbf{u}_{\mathrm{cl}}^{n+1} = \mathbf{v}_{\mathrm{cl}}^{n+\frac{1}{2}} \tag{6.49}$$

Note that the RHS of (6.46–6.49) are matrix-operator-free, while the LHS will result in tridiagonal matrices that can be solved efficiently. Alternative to ADI, one may also resort to the multiple LOD CL-FDTD method using the proper split matrices for stability and efficiency [71–73].

6.3 MOBILE APPS FOR EM EDUCATION

Exploiting the wide availability of mobile devices, several educational mobile apps have been developed for enhanced teaching and learning of EM, e.g., *EMpolarization*, *EMwaveRT*, *MuStripKit*, etc. (some are available on App/Play Store) [23–29]. These apps are incorporated with innovative CEM methods that run efficiently on mobile devices such as smartphones or iPads and may be supplemented with 3-D displays. They provide real-time EM and circuit simulations as well as 2-D and 3-D visualizations of wave phenomena. Several short video clips demonstrating some of the mobile apps are available from [74].

6.3.1 ELECTROMAGNETIC POLARIZATION

Let us first consider the teaching and learning of electromagnetic polarization. Figure 6.1 shows the illustrations for the loci of electric field vector tip at successive time instants for (a) linear, (b) circular, and (c) elliptical polarizations. Since the illustrations appear static only in textbooks or lecture notes, one relies on the indication of time instants t_1, t_2, ..., t_5, etc., to trace out various polarization loci and handedness in a rather laborious manner. Furthermore, such illustrations are fixed with preset parameters and they are not configurable interactively.

To enhance the teaching and learning, we make use of *EMpolarization* mobile app to provide 2-D and 3-D visualizations of polarizations [23–25]. Figure 6.2 shows the snapshots of (a) electric field vector decomposition and polarization magnitude-phase conditions, as well as (b) polarization ellipse, Stokes parameters, and Poincare sphere. The app provides interactive animations of various polarizations that are configurable in real time based on user input. One can use the slider or on-screen keyboard to change incrementally or enter specific values of polarization parameters. The field animations can be paused or continued as needed, with the 3-D view being zoomable or rotatable in any desired direction. These help students identify clearly the polarization type, such as linear, circular, and elliptical polarization, as well as the handedness.

More in-depth calculations of various polarization ellipse angles and Stokes parameters are available on the mobile app. Students may refer them for self-learning of these topics or checking their own calculated answers. For the Poincare sphere on the app, students could even have fun rolling the red ball on the earth globe in Figure 6.2(b). They can be engaged to work out the polarization state corresponding to any place of interest, e.g., linear polarization at Ecuador, left-hand elliptical polarization at Hawaii, right-hand circular polarization at the South Pole, etc.

6.3.2 PLANE WAVE REFLECTION AND TRANSMISSION

Let us next consider the teaching and learning of plane wave reflection and transmission. Figure 6.3 shows the illustrations of plane wave reflection and transmission for

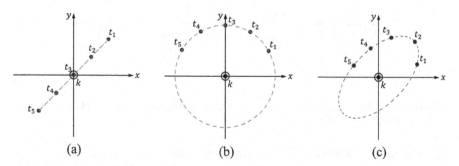

(a) (b) (c)

FIGURE 6.1 Illustrations for the loci of electric field vector tip at successive time instants for (a) linear, (b) circular, and (c) elliptical polarizations.

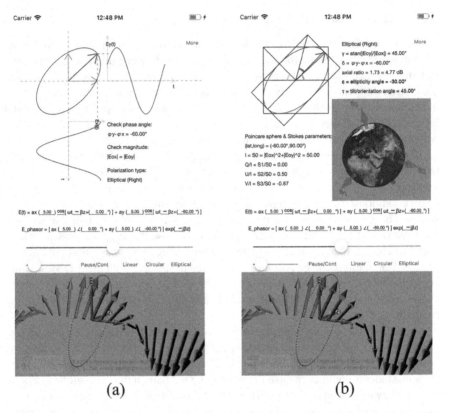

FIGURE 6.2 *EMpolarization* mobile app snapshots of (a) electric field vector decomposition and polarization magnitude-phase conditions; (b) polarization ellipse, Poincare sphere and Stokes parameters.

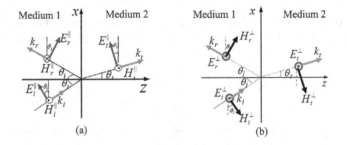

FIGURE 6.3 Illustrations of plane wave reflection and transmission for oblique incidence with (a) parallel and (b) perpendicular polarizations.

oblique incidence with (a) parallel and (b) perpendicular polarizations. Such illustrations in textbooks or lecture notes can depict the incident, reflected, and transmitted waves in static 2-D view only. To enhance the teaching and learning, we make use of *EMwaveRT* mobile app supplementable with 3-D TV [26,27]. Figure 6.4 shows the snapshots of plane wave reflection and transmission for oblique incidence with (a)

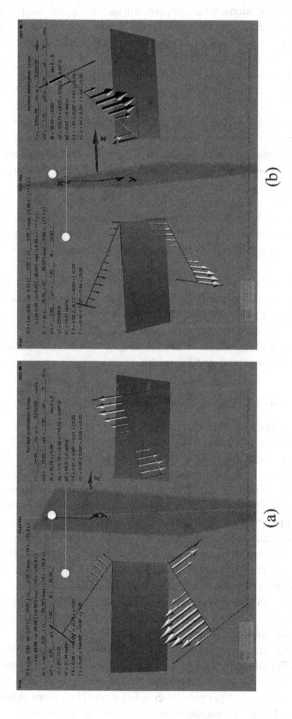

FIGURE 6.4 *EMwaveRT* mobile app snapshots of plane wave reflection and transmission for oblique incidence with (a) parallel and (b) perpendicular polarizations.

parallel and (b) perpendicular polarizations. The app depicts the incident, reflected, and transmitted waves, along with the plane of incidence and the planar interface between media 1 and 2. All these are in 3-D views, which can be rotated and zoomed in/out flexibly. For general oblique incidence and polarization based on user input, the field animations can be visualized in real time.

The plane wave reflection and transmission app deals with many plane wave and material parameters. These parameters are user-configurable and can be shown or hidden as and when needed. There are also field tangential components of incident, reflected, and transmitted waves, which can be shown or hidden to be projected onto the interface plane. When shown, one can observe real-time visualization of matching them across the interface at every instant. This helps students see and appreciate the boundary conditions in action, rather than in plain static expressions as in textbooks. The mobile app facilitates the visualization of various EM wave reflection and transmission phenomena. Starting from the normal incidence and by increasing the incident angle, the evolutions of wave polarization and propagation can be animated. These include the interesting cases of the Brewster phenomenon showing linearly polarized reflected wave, as well as the total internal reflection at or beyond the critical angle.

To further enhance visualization, 3-D displays such as 3-D TV may be used [26]. The 3-D TV program that is used to generate 3-D animations is available for download from [75]. Upon projecting the animations onto 3-D TV, one can observe the wave interactions in stereoscopic view with depth perception using 3-D glasses. Using demonstrations via 3-D TV in classroom or at home, students could actually watch "interactive 3-D movie" and learn EM!

6.3.3 TRANSMISSION LINES AND COUPLED-LINE STRUCTURES

We have also developed various mobile apps for transmission lines, stubs, coupled lines, couplers, filters, and more. These mobile apps are useful to aid teaching and learning of microstrip circuits and transmission line concepts such as standing waves, mismatch reflections, impedance matching, resonances, etc. [28–32]. Figure 6.5 shows the *MuStripKit* mobile app snapshots of (a) microstrip configuration, (b) circuit schematic construction, and (c) S parameters calculation. The mobile app allows users to construct practical microstrip circuit and input circuit parameters with touch-based interactivity. Such user-configurable circuit may comprise transmission lines, open- and short-circuited stubs, as well as lumped elements such as resistors, capacitors, and inductors in parallel and/or series. Using the M1-D FDTD method described in the previous section, the app can provide real-time simulations and visualizations of EM wave propagation on mobile devices. However, due to the stability constraint on the time step size, one may need to wait long for observing the EM dynamics (e.g., reflections) on long transmission lines.

To overcome the stability constraint, we have resorted to unconditionally stable M1-D FADI-FDTD allowing larger CFLN in some mobile apps. Figure 6.6 shows the mobile app snapshots of (a) transmission line with open- and short-circuited stubs, (b) coupled lines, and (c) parallel coupled filter. These apps provide efficient simulations of EM wave propagation and coupling mechanisms on transmission

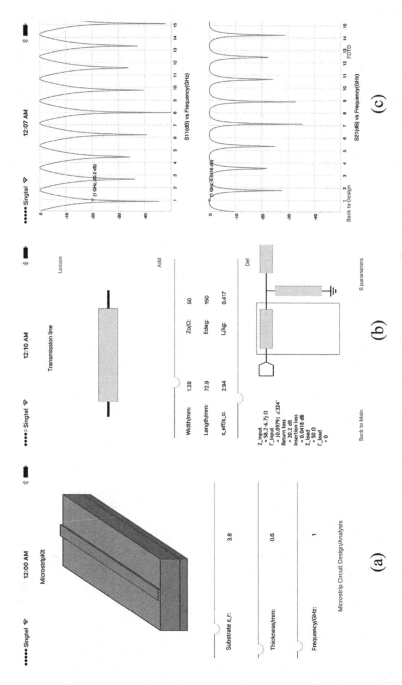

FIGURE 6.5 *MuStripKit* mobile app snapshots of (a) microstrip configuration, (b) circuit schematic construction, and (c) S parameters calculation.

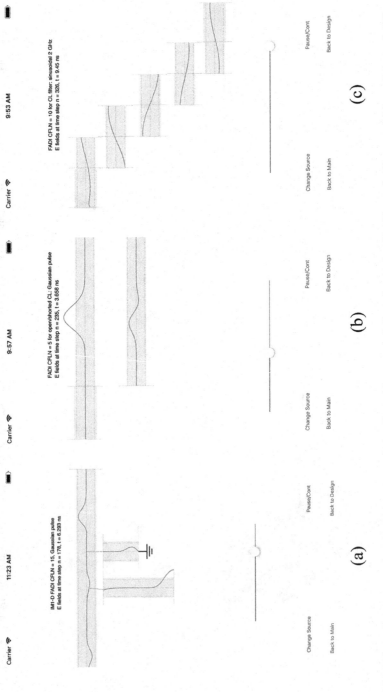

FIGURE 6.6 Mobile app snapshots of (a) transmission line with open- and short-circuited stubs, (b) coupled lines, and (c) parallel coupled filter.

lines and coupled-line structures using CFLN > 1. For instance, one can quickly observe the dynamics of EM pulse being reflected from the open-/short-circuited stub ends, and coupled from line 1 to line 2 in Figure 6.6(a) and (b), respectively. For the parallel coupled filter with input sinusoid within passband, one can see how the waves resonate constructively while being coupled through in Figure 6.6(c). Using M1-D FADI-FDTD with larger CFLN, the simulations are "fast-forwardable" to provide insightful visualizations of propagation, resonating and coupling mechanisms.

6.3.4 EDUCATIONAL STUDY AND SURVEY RESULTS

Our pedagogy approach includes demonstration and hands-on sessions using mobile apps and 3-D display during EM teaching and learning. It should be emphasized here that such approach is not meant to fully replace all existing EM teaching practices with PC-based hands-on and lab sessions, which still play an important role by providing hardware and software exposures. Instead, it is intended to be integrated into the current practices with the aim for enhancing EM teaching and learning while providing a more holistic approach. Using the mobile apps and 3-D display, our educational study included feedback and assessment to evaluate their effectiveness and for continuous improvement of the contents. In particular, we carried out the study during undergraduate classes of EE3001 Engineering Electromagnetics and Master of Science (MSc) classes of EE6128 RF Circuits for Wireless Communications, at the School of Electrical and Electronic Engineering, Nanyang Technological University, Singapore. Students were given pre- and post-tests along with some surveys before and after using mobile apps/3-D TV. Some key findings and results from the tests and surveys conducted on various EM topics are discussed in the following.

On the topic of EM wave polarization using *EMpolarization* mobile app, it has been found that out of 106 undergraduate students [24,25],

- 66% of them found that the topic related to polarization is not easy to understand;
- 69% of them thought that the app motivates them to learn more actively;
- 87% of them found that the app is interesting and intriguing;
- 78% of them thought that their understanding of the wave polarization has been improved after being taught with the aid of the app.

On the topic of plane wave reflection and transmission using 3-D TV and *EMwaveRT* mobile app, it has been found that out of 58 undergraduate students [26],

- 89% of them agreed strongly or moderately that the 3-D TV provides depth sensation to help them identify easily the direction and position of various E field vectors;
- they provided an average rating of 8 out of 10 when asked to rate the 3-D TV in helping them visualize and understand the reflection and transmission of plane waves;

- when tested on the polarization type of the reflected wave at Brewster angle, 31% of them answered correctly as linear polarization and 26% answered correctly as perpendicular component during pre-test;
- 93% of them answered correctly as linear polarization and perpendicular component during post-test.

On the topic of transmission lines using *MuStripKit* mobile app, it has been found that out of 63 undergraduate students [31],

- 98% of them agreed strongly or moderately that the app is interesting and motivates them to learn more actively on the topics of transmission lines;
- they indicated with an average rating of 8.6 out of 10 that the app helped them understand wave propagation on transmission lines;
- when asked to sketch the standing wave pattern on transmission line terminated with short- or open-circuited load and whether the standing wave is still oscillating, 43% of them sketched the pattern correctly and 41% answered correctly as oscillating during pre-test;
- 94% of them sketched the pattern correctly and 95% answered correctly as oscillating during post-test.

The topics of transmission lines, coupled lines, and couplers are applicable for advanced graduate-level microwave circuit course. Using the corresponding mobile apps, it has been found that out of 74 MSc students [32],

- 97% of them agreed strongly or moderately that the apps are interesting and motivate them to learn more actively on the topics of transmission lines, coupled lines, and couplers;
- they provided an average rating of 9.3 out of 10 when asked to rate the apps in helping them understand the concepts of transmission lines.

Overall there is an improvement in the students' understanding after using mobile apps and 3-D TV. The feedback results generally show good responses from the students.

6.3.5 APPLICATIONS AND EXTENSIONS OF MOBILE APPS

With the mobile apps incorporating efficient CEM methods, real-time simulations can be carried out for quick design, analysis, teaching, and learning. The mobile apps can provide ubiquitous interactive simulations and insightful visualizations for students anytime, anywhere. Using educational apps on handy mobile devices bypasses the hassle of accessing high-performance workstation and computing lab software (especially for large classes). During the COVID-19 pandemic lockdown and school closures, there was no physical access to classrooms or labs, while some software servers were suspended or operated with limited capacity. The mobile apps have been the key enablers providing seamless interactive simulations and insightful visualizations even at home. Students could continue with hands-on design and analysis using the mobile toolkit apps to work on their homework and assignments [26,28].

The mobile apps with touch-based interactivity and 2-D/3-D visualization features can facilitate students' self-learning and further exploration. For instance, while viewing the demonstration of positive refraction in usual double-positive medium, some may be interested to observe negative refraction in double-negative medium, or even query about refraction in single-negative medium, etc. To explore all these, one just need to drag the slider or enter the parameter values to visualize the corresponding plane wave reflection and transmission phenomena on the mobile app [27]. More apps with advanced CEM can be developed to determine and demonstrate EM waves in multilayered complex media. They can also be enhanced with artificial intelligence (AI) via innovative mobile intelligent tutoring systems, along with thorough and rigorous assessments incorporated into online/mobile platforms [33]. Such mobile apps would be useful not only for interactive teaching and learning, but also for tutoring, assessments and even paperless examinations.

6.4 CONCLUSION

This chapter has presented the CEM methods and mobile apps for EM education. Some CEM methods in time domain that provide much intuitional insight for wave propagations and dynamics have been reviewed. These include the conditionally stable explicit 3-D FDTD method and unconditionally stable implicit 3-D FDTD methods. For the latter, the FADI and FLOD FDTD methods have been highlighted for their concise, simple, and efficient update equations with matrix-operator-free RHS. Further developments using more efficient explicit and implicit M1-D FDTD and CL-FDTD methods have also been discussed. Several educational mobile apps have been described for enhanced teaching and learning of EM on various topics including EM polarization, plane wave reflection and transmission, microstrip circuits, transmission lines, and coupled-line structures, etc. These mobile apps feature convenient and effective touch-based interactivity to aid teaching and learning. They are incorporated with innovative CEM methods that are efficient and fast-forwardable using unconditionally stable implicit methods. Interactive simulations and insightful visualizations of wave propagation are provided on the mobile devices to elucidate the concepts of EM theory, transmission lines, and microwave circuits. Besides mobile devices, 3-D displays such as 3-D TV may be supplemented for enhanced EM education as well. Some educational study and survey results using mobile devices/3-D TV have been discussed, which show an improvement in students' understanding and good responses. It is hoped that the present chapter would motivate and inspire further research and applications of CEM and mobile apps for education.

BIBLIOGRAPHY

1. M. F. Iskander, "Computer-based electromagnetic education," *IEEE Trans. Microw. Theory Tech.*, vol. 41, no. 6, pp. 920–931, 1993.
2. M. Munoz, "Using graphical based software to aid in the understanding of electromagnetic field theory," *Proc. Frontiers in Education Conf.*, Salt Lake City, UT, pp. 863–867, 1996.

3. K. Preis, I. Bardi, O. Biro, R. Hoschek, M. Mayr and I. Ticar, "A virtual electromagnetic laboratory for the classroom and the WWW," *IEEE Trans. Magnetics*, vol. 33, no. 2, pp. 1990–1993, 1997.

4. B. Beker, D. W. Bailey and G. J. Cokkinides, "An application-enhanced approach to introductory electromagnetics," *IEEE Trans. Education*, vol. 41, no. 1, pp. 31–36, 1998.

5. E. Okayama, V. Cingoski, S. Noguchi, K. Kaneda and H. Yamashita, "Interactive visualization system for education and design in electromagnetics," *IEEE Trans. Magnetics*, vol. 36, no. 4, pp. 995–999, 2000.

6. M. F. Iskander, "Technology-based electromagnetic education," *IEEE Trans. Microw. Theory Tech.*, vol. 50, no. 3, pp. 1015–1020, 2002.

7. K. W. E. Cheng, X. D. Xue, K. F. Kwok and T. K. Cheung, "Improvement of classroom teaching of electromagnetics by means of an electronic book," *IEEE Trans. Magnetics*, vol. 39, no. 3, pp. 1570–1573, 2003.

8. W. J. R. Hoefer and P. P. M. So, "A time-domain virtual electromagnetics laboratory for microwave engineering education," *IEEE Trans. Microw. Theory Tech.*, vol. 51, no. 4, pp. 1318–1325, 2003.

9. M. de Magistris, "A MATLAB-based virtual laboratory for teaching introductory quasi-stationary electromagnetics," *IEEE Trans. Education*, vol. 48, no. 1, pp. 81–88, 2005.

10. V. Pulijala, A. R. Akula and A. Syed, "A Web-based virtual laboratory for electromagnetic theory," *IEEE Int. Conf. Technology for Education*, Kharagpur, pp. 13–18, 2013.

11. M. F. Iskander, "Multimedia and technology based electromagnetic education," *IEEE AP-S Int. Symp. Antennas Propag.*, Memphis, TN, pp. 531–532, 2014.

12. A. Taflove and S. C. Hagness, *Computational Electrodynamics: The Finite-Difference Time-Domain Method*, Boston, M. A.: Artech House, 2005.

13. K. S. Yee, "Numerical solution of initial boundary value problems involving Maxwell's equations in isotropic media," *IEEE Trans. Antennas Propag.*, Vol. 14, No. 3, pp. 302–307, May 1966.

14. F. Zheng, Z. Chen and J. Zhang, "Toward the development of a three-dimensional unconditionally stable finite-difference time-domain method," *IEEE Trans. Microw. Theory Tech.*, Vol. 48, No. 9, pp. 1550–1558, Sep. 2000.

15. T. Namiki, "3-D ADI-FDTD method – Unconditionally stable time-domain algorithm for solving full vector Maxwell's equations," *IEEE Trans. Microw. Theory Tech.*, Vol. 48, No. 10, pp. 1743–1748, Oct. 2000.

16. W. Fu and E. L. Tan, "Development of split-step FDTD method with higher order spatial accuracy," *Electron. Lett.*, Vol. 40, No. 20, pp. 1252–1254, Sep. 2004.

17. W. Fu and E. L. Tan, "Compact higher-order split-step FDTD method," *Electron. Lett.*, Vol. 41, No. 7, pp. 397–399, Mar. 2005.

18. E. L. Tan, "Unconditionally stable LOD-FDTD method for 3-D Maxwell's equations," *IEEE Microw. Wireless Compon. Lett.*, Vol. 17, No. 2, pp. 85–87, Feb. 2007.

19. E. L. Tan, "Fundamental schemes for efficient unconditionally stable implicit finite-difference time-domain methods," *IEEE Trans. Antennas Propag.*, Vol. 56, No. 1, pp. 170–177, Jan. 2008.

20. E. L. Tan, "Efficient algorithm for the unconditionally stable 3-D ADI-FDTD method," *IEEE Microw. Wireless Compon. Lett.*, Vol. 17, No. 1, pp. 7–9, Jan. 2007.

21. E. L. Tan, "Concise current source implementation for efficient 3-D ADI-FDTD method," *IEEE Microw. Wireless Compon. Lett.*, Vol. 17, No. 11, pp. 748–750, Nov. 2007.

22. T. H. Gan and E. L. Tan, "Current source implementations for fundamental SS2-FDTD method," *Asia-Pacific Microwave Conf.*, Kaohsiung, Taiwan, pp. 1292–1294, Dec. 2012.

23. E. L. Tan and D. Y. Heh, "Demonstration of electromagnetic polarization app on iPad," *IEEE Int. Conf. Comput. Electromagn.*, Kumamoto, Japan, pp. 196–197, Mar. 2017.

24. E. L. Tan and D. Y. Heh, "Mobile device aided teaching and learning of electromagnetic polarization," *IEEE 6th Int. Conf. Teaching, Assessment, and Learning for Engineering*, Hong Kong, pp. 52–55, Dec. 2017.

25. E. L. Tan and D. Y. Heh, "Teaching and learning electromagnetic polarization using mobile devices," *IEEE Antennas Propag. Mag.*, Vol. 60, No. 4, pp. 112–121, Aug. 2018.

26. E. L. Tan and D. Y. Heh, "Teaching and learning electromagnetic plane wave reflection and transmission using 3-D TV," *IEEE Antennas Propag. Mag.*, Vol. 61, No. 2, pp. 101–108, Apr. 2019.

27. E. L. Tan and D. Y. Heh, "Demonstration of electromagnetic plane wave reflection and transmission on iPad," *IEEE Int. Conf. Comput. Electromagn.*, Singapore, Aug. 2020.

28. Z. Yang and E. L. Tan, "A microstrip circuit tool kit app with FDTD analysis including lumped elements," *IEEE Microw. Mag.*, Vol. 16, No. 1, pp. 74–80, 2015.

29. Z. Yang and E. L. Tan, "A microwave transmission line courseware based on multiple 1-D FDTD method on mobile devices," *Asia-Pacific Conf. Antennas Propag.*, Bali, Indonesia, pp. 251–252, 2015.

30. E. L. Tan and D. Y. Heh, "Demonstration of electromagnetic waves propagation along transmission lines on iPad," *2018 Joint IEEE Int. Symp. Electromag. Compat. and Asia-Pacific Symp. Electromag. Compat.*, Singapore, pp. 599–601, May 2018.

31. E. L. Tan and D. Y. Heh, "M1-D FDTD methods for mobile interactive teaching and learning of wave propagation in transmission lines," *IEEE Antennas Propag. Mag.*, Vol. 61, No. 5, pp. 119–126, Oct. 2019.

32. E. L. Tan and D. Y. Heh, "Mobile teaching and learning of coupled-line structures: The multiple-1D coupled-line finite-difference time-domain method," *IEEE Antennas Propag. Mag.*, Vol. 62, No. 4, pp. 62–69, Apr. 2020.

33. E. L. Tan, "Fundamental implicit FDTD schemes for computational electromagnetics and educational mobile apps (Invited review)," *Progress In Electromagnetics Research*, Vol. 168, pp. 39–59, 2020.

34. E. L. Tan, "Efficient algorithms for Crank-Nicolson-based finite-difference time-domain methods," *IEEE Trans. Microw. Theory Tech.*, Vol. 56, No. 2, pp. 408–413, Feb. 2008.

35. J. Shibayama, T. Hirano, J. Yamauchi and H. Nakano "Efficient implementation of frequency-dependent 3-D LOD-FDTD method using fundamental scheme," *Electron. Lett.*, Vol. 48, No. 13, pp. 774–775, Jun. 2012.

36. T. H. Gan and E. L. Tan, "Unconditionally stable fundamental LOD-FDTD method with second-order temporal accuracy and complying divergence," *IEEE Trans. Antennas Propag.*, Vol. 61, No. 5, pp. 2630–2638, May 2013.

37. T. H. Gan and E. L. Tan, "Convolutional perfectly matched layer (CPML) for fundamental LOD-FDTD method with 2nd order temporal accuracy and complying divergence," *Asia-Pacific Microwave Conf.*, Seoul, Korea, pp. 839–841, Nov. 2013.

38. T. H. Gan and E. L. Tan, "Application of the fundamental LOD2-CD-FDTD method for antenna modeling," *Asia-Pacific Conf. Antennas Propag.*, Bali, Indonesia, pp. 445–446, 2015.

39. W. Fu and E. L. Tan, "A compact higher-order ADI-FDTD method," *Microwave Opt. Technol. Lett.*, Vol. 44, No. 3, pp. 273–275, Feb. 2005.

40. W. Fu and E. L. Tan, "A parameter optimized ADI-FDTD method based on the (2,4) stencil," *IEEE Trans. Antennas Propag.*, Vol. 54, No. 6, pp. 1836–1842, Jun. 2006.

41. D. Y. Heh and E. L. Tan, "Efficient implementation of 3-D ADI-FDTD method for lossy media," *IEEE MTT-S Int. Microwave Symp.*, Boston, Massachusetts, pp. 313–316, Jun. 2009.

42. D. Y. Heh and E. L. Tan, "Unified efficient fundamental ADI-FDTD schemes for lossy media," *Progress In Electromagnetics Research B*, Vol. 32, pp. 217–242, Jul. 2011.

43. D. Y. Heh and E. L. Tan, "Stable formulation of FADI-FDTD method for multiterm, doubly, second-order dispersive media," *IEEE Trans. Antennas Propag.*, Vol. 61, No. 8, pp. 4167–4175, Aug. 2013.

44. W. C. Tay and E. L. Tan, "Implementation of Mur first order absorbing boundary condition in efficient 3-D ADI-FDTD," *IEEE AP-S Int. Symp. Antennas Propag.*, Charleston, South Carolina, USA, Jun. 2009.

45. W. C. Tay and E. L. Tan, "Split-field PML implementation for the efficient fundamental ADI-FDTD method," *Asia-Pacific Microwave Conf.*, Singapore, pp. 1553–1556, Dec. 2009.

46. E. L. Tan, "Acceleration of LOD-FDTD method using fundamental scheme on graphics processor units," *IEEE Microw. Wireless Compon. Lett.*, Vol. 20, No. 12, pp. 648–650, Dec. 2010.

47. W. C. Tay, D. Y. Heh and E. L. Tan, "GPU-accelerated fundamental ADI-FDTD with complex frequency shifted convolutional perfectly matched layer," *Progress In Electromagnetics Research M*, Vol. 14, pp. 177–192, Oct. 2010.

48. W. C. Tay and E. L. Tan, "Mur absorbing condition for efficient fundamental 3-D LOD-FDTD," *IEEE Microw. Wireless Compon. Lett.*, Vol. 20, No. 2, pp. 61–63, Feb. 2010.

49. W. C. Tay and E. L. Tan, "Implementations of PMC and PEC boundary conditions for efficient fundamental ADI and LOD-FDTD," *J. Electrom. Waves Applicat.*, Vol. 24, No. 4, pp. 565–573, Mar. 2010.

50. G. Singh, E. L. Tan and Z. N. Chen, "A split-step FDTD method for 3-D Maxwell's equations in general anisotropic media," *IEEE Trans. Antennas Propag.*, Vol. 58, No. 11, pp. 3647–3657, Nov. 2010.

51. Z. Yang and E. L. Tan, "3-D non-uniform time step locally one-dimensional FDTD method," *Electron. Lett.*, Vol. 52, No. 12, pp. 993–994, Jun. 2016.

52. Z. Yang and E. L. Tan, "Stability analyses of non-uniform time-step schemes for ADI- and LOD-FDTD methods," *IEEE Int. Conf. Comput. Electromagn.*, Kumamoto, Japan, pp. 312–313, Mar. 2017.

53. E. L. Tan and D. Y. Heh, "Stability analyses of nonuniform time-step LOD-FDTD methods for electromagnetic and thermal simulations," *IEEE J. Multiscale Multiphys. Comput. Tech.*, Vol. 2, pp. 183–193, Nov. 2017.

54. Z. Yang, E. L. Tan and D. Y. Heh, "Second-order temporal-accurate scheme for 3-D LOD-FDTD method with three split matrices," *IEEE Microw. Wireless Compon. Lett.*, Vol. 14, pp. 1105–1108, Jan. 2015.

55. Z. Yang, E. L. Tan and D. Y. Heh, "Variants of second-order temporal-accurate 3-D FLOD-FDTD schemes with three split matrices," *IEEE Int. Conf. Comput. Electromagn.*, Guangzhou, China, pp. 265–267, Feb. 2016.

56. Z. Yang and E. L. Tan, "Multiple one-dimensional FDTD method for coupled transmission lines and stability condition," *IEEE Microw. Wireless Compon. Lett.*, Vol. 26, No. 11, pp. 864–866, Nov. 2016.

57. Z. Yang and E. L. Tan, "Multiple one-dimensional finite-difference time-domain method for asymmetric coupled transmission lines," *IEEE Int. Conf. Comput. Electromagn.*, Chengdu, China, Mar. 2018.

58. C. R. Paul, *Analysis of Multiconductor Transmission Lines*, 2nd ed., New York, NY, USA: Wiley, 2008.

59. R. K. Mongia, I. J. Bahl, P. Bhartia, and J. Hong, *RF and Microwave Coupled-Line Circuits*, 2nd ed., Norwood, MA, USA: Artech House, 2007.

60. D. M. Pozar, *Microwave Engineering*, 4th ed., New York: Wiley, 2011.

61. Z. Yang and E. L. Tan, "Interconnected multi-1-D FADI- and FLOD-FDTD methods for transmission lines with interjunctions," *IEEE Trans. Microw. Theory Tech.*, Vol. 65, No. 3, pp. 684–692, Mar. 2017.

62. E. L. Tan and Z. Yang, "Non-uniform time-step FLOD-FDTD method for multiconductor transmission lines including lumped elements," *IEEE Trans. Electromagn. Compat.*, Vol. 59, No. 6, pp. 1983–1992, Jun. 2017.

63. W. Fu and E. L. Tan, "ADI-FDTD method including linear lumped networks," *Electron. Lett.*, Vol. 42, No. 13, pp. 728–729, Jun. 2006.

64. W. Fu and E. L. Tan, "Unconditionally stable ADI-FDTD method including passive lumped elements," *IEEE Trans. Electromagn. Compat.*, Vol. 48, No. 4, pp. 661–668, Nov. 2006.

65. Z. Yang and E. L. Tan, "Efficient 3-D fundamental LOD-FDTD method with lumped elements," *IEEE MTT-S Int. Microwave Symp.*, Tampa, Florida, USA, Jun. 2014.

66. Z. Yang and E. L. Tan, "3-D unified FLOD-FDTD method incorporated with lumped elements," *IEEE AP-S Int. Symp. Antennas Propag.*, San Diego, USA, 9-14 July 2017.

67. D. Y. Heh and E. L. Tan, "Unconditionally stable multiple one-dimensional ADI-FDTD method for coupled transmission lines," *IEEE Trans. Antennas Propag.*, Vol. 66, No. 12, pp. 7488–7492, Dec. 2018.

68. D. Y. Heh and E. L. Tan, "Numerical stability analysis of M1-D ADI-FDTD method for coupled transmission lines," *IEEE Int. Conf. Comput. Electromagn.*, Shanghai, China, Mar. 2019.

69. E. L. Tan and D. Y. Heh, "Source-incorporated M1-D FADI-FDTD method for coupled transmission lines," *11th Int. Conf. Microw. Millimeter Wave Techn.*, Guangzhou, China, May 2019.

70. E. L. Tan and D. Y. Heh, "Multiple 1-D fundamental ADI-FDTD method for coupled transmission lines on mobile devices," *IEEE J. Multiscale Multiphys. Comput. Tech.*, Vol. 4, No. 1, pp. 198–206, Dec. 2019.

71. D. Y. Heh and E. L. Tan, "Numerical stability analysis of M1-D LOD-FDTD method for inhomogeneous coupled transmission lines," *IEEE AP-S Int. Symp. Antennas Propag.*, Atlanta, USA, pp. 1657–1658, Jul. 2019.

72. D. Y. Heh and E. L. Tan, "Multiple LOD-FDTD method for inhomogeneous coupled transmission lines and stability analyses," *IEEE Trans. Antennas Propag.*, Vol. 68, No. 3, pp. 2198–2205, Mar. 2020.

73. D. Y. Heh and E. L. Tan, "Multiple LOD-FDTD method for multiconductor coupled transmission lines," *IEEE J. Multiscale Multiphys. Comput. Tech.*, Vol. 5, pp. 201–208, Sep. 2020.

74. E. L. Tan, "Mobile apps with CEM for teaching and learning of electromagnetics," [Online]. Available at: https://personal.ntu.edu.sg/eeltan/TELmobileapps.html (Also accessible from the IEEE AP-S Resource Center at https://resourcecenter.ieeeaps.org/).

75. E. L. Tan, "Download page for 3-D TV program," [Online]. Available at: https://personal.ntu.edu.sg/eeltan/TEL3DTV.html

7 Teaching Electromagnetic Field Theory Using Differential Forms

Karl F. Warnick

Brigham Young University, USA

CONTENTS

7.1   INTRODUCTION

Certain questions are often asked by students of electromagnetic (EM) field theory: Why does one need both field intensity and flux density to describe a single field? How does one visualize the curl operation? Is there some way to make Ampere's law or Faraday's law as physically intuitive as Gauss's law? The Stokes theorem and the divergence theorem seem vaguely similar; do they have a deeper connection? Because of difficulty with concepts related to these questions, some students leave introductory courses lacking a real understanding of the physics of electromagnetics. Interestingly, none of these concepts are intrinsically more difficult than other aspects of EM theory; rather, they are unclear because of the limitations of the mathematical language traditionally used to teach electromagnetics: vector analysis. We show that the calculus of differential forms clarifies these and other fundamental principles of electromagnetic field theory.

The use of the calculus of differential forms in electromagnetics has been explored in several important papers and texts, including Misner, Thorne, and Wheeler [3], Deschamps [4], and Burke [2]. These works note some of the advantages of the use of differential forms in EM theory. Misner *et al.* and Burke treat the graphical representation of forms and operations on forms, as well as other aspects of the application of forms to electromagnetics. Deschamps was among the first to advocate the use of forms in teaching engineering electromagnetics.

Existing treatments of differential forms in EM theory either target an advanced audience or are not intended to provide a complete exposition of the pedagogical advantages of differential forms. This chapter presents the topic on an undergraduate level and emphasizes the benefits of differential forms in teaching introductory electromagnetics, especially graphical representations of forms and operators. The calculus of differential forms and principles of EM theory are introduced in parallel, much as would be done in a beginning EM course. We present concrete visual pictures of the various field quantities, Maxwell's laws, and boundary conditions. The aim is to demonstrate that differential forms are an attractive and viable alternative to vector analysis as a tool for teaching electromagnetic field theory.

7.1.1 DEVELOPMENT OF DIFFERENTIAL FORMS

Cartan and others developed the calculus of differential forms in the early 1900s. A differential form is a quantity that can be integrated, including differentials. More precisely, a differential form is a fully covariant, fully antisymmetric tensor. The calculus of differential forms is a self–contained subset of tensor analysis.

Since Cartan's time, the use of forms has spread to many fields of pure and applied mathematics, from differential topology to the theory of differential equations. Differential forms are used by physicists in general relativity [3], quantum field theory [5], thermodynamics [6], mechanics [7], as well as electromagnetics. A section on differential forms is commonplace in mathematical physics texts [8,9]. Differential forms have been applied to control theory by Hermann [10] and others.

7.1.2 DIFFERENTIAL FORMS IN EM THEORY

The laws of electromagnetic field theory as expressed by James Clerk Maxwell in the mid-1800s required dozens of equations. Vector analysis offered a more convenient tool for working with EM theory than earlier methods. Tensor analysis is in turn more concise and general, but is too abstract to give students a conceptual understanding of EM theory. Weyl and Poincaré expressed Maxwell's laws using differential forms early this century. Applied to electromagnetics, differential forms combine much of the generality of tensors with the simplicity and concreteness of vectors.

General treatments of differential forms and EM theory include papers [4,11–15]. Ingarden and Jamiolkowksi [16] is an electrodynamics text using a mix of vectors and differential forms. Parrott [17] employs differential forms to treat advanced electrodynamics. Thirring [18] is a classical field theory text that includes certain applied topics such as waveguides. Bamberg and Sternberg [6] develop a range of topics in mathematical physics, including EM theory via a discussion of discrete forms and circuit theory. Burke [2] treats a range of physics topics using forms, shows how to graphically represent forms, and gives a useful discussion of twisted differential forms. The general relativity text by Misner, Thorne and Wheeler [3] has several chapters on EM theory and differential forms, emphasizing the graphical representation of forms. Flanders [7] treats the calculus of forms and various applications, briefly mentioning electromagnetics.

We note here that many authors, including most of those referenced above, give the spacetime formulation of Maxwell's laws using forms, in which time is included as a differential. We use only the (3+1) representation, since the spacetime representation is treated in many references and is not as convenient for various elementary and applied topics. Other formalisms for EM theory are available, including bivectors, quaternions, spinors, and higher Clifford algebras. None of these offer the combination of concrete graphical representations, ease of presentation, and close relationship to traditional vector methods that the calculus of differential forms brings to undergraduate-level electromagnetics.

The tools of applied electromagnetics have begun to be reformulated using differential forms. The authors have developed a convenient representation of electromagnetic boundary conditions (7.20). Thirring [18] treats several applications of EM theory using forms.

7.1.3 PEDAGOGICAL ADVANTAGES OF DIFFERENTIAL FORMS

As a language for teaching electromagnetics, differential forms offer several important advantages over vector analysis. Vector analysis allows only two types of quantities: scalar fields and vector fields (ignoring inversion properties). In a three–dimensional space, differential forms of four different types are available. This allows flux density and field intensity to have distinct mathematical expressions and graphical representations, providing the student with mental pictures that clearly reveal the different properties of each type of quantity. The physical interpretation of a vector field is often implicitly contained in the choice of operator or integral that acts on it. With differential forms, these properties are directly evident in the type of form used to represent the quantity.

The basic derivative operators of vector analysis are the gradient, curl, and divergence. The gradient and divergence lend themselves readily to geometric interpretation, but the curl operation is more difficult to visualize. The gradient, curl, and divergence become special cases of a single operator, the exterior derivative, and the curl obtains a graphical representation that is as clear as that for the divergence. The physical meanings of the curl operation and the integral expressions of Faraday's and Ampere's laws become so intuitive that the usual order of development can be reversed by introducing Faraday's and Ampere's laws to students first and using these to motivate Gauss's laws.

The Stokes theorem and the divergence theorem have an obvious connection in that they relate integrals over a boundary to integrals over the region inside the boundary, but in the language of vector analysis they appear very different. These theorems are special cases of the generalized Stokes theorem for differential forms, which also has a simple graphical interpretation.

Since 1992, in the Brigham Young University Department of Electrical and Computer Engineering we have incorporated short segments on differential forms into our beginning, intermediate, and graduate electromagnetics courses. In the Fall of 1995, we reworked the entire beginning electromagnetics course, changing emphasis from vector analysis to differential forms. Following the first semester in which the new curriculum was used, students completed a detailed written evaluation. Out of 44 responses, four were partially negative; the rest were in favor of the change to differential forms. Certainly, enthusiasm of students involved in something new increased the likelihood of positive responses, but one fact was clear: pictures of differential forms helped students understand the principles of electromagnetics.

7.2 DIFFERENTIAL FORMS AND THE ELECTROMAGNETIC FIELD

In this section, we define differential forms of various degrees and identify them with field intensity, flux density, current density, charge density, and scalar potential.

A differential form is a quantity that can be integrated, including differentials. $3x\,dx$ is a differential form, as are $x^2y\,dx\,dy$ and $f(x,y,z)\,dy\,dz + g(x,y,z)\,dz\,dx$. The type of integral called for by a differential form determines its degree. The form $3x\,dx$ is integrated under a single integral over a path and so is a 1-form. The form $x^2y\,dx\,dy$ is integrated by a double integral over a surface, so its degree is two. A 3-form is integrated by a triple integral over a volume. 0-forms are functions, "integrated" by evaluation at a point. Table 7.1 gives examples of forms of various degrees. The coefficients of the forms can be functions of position, time, and other variables.

7.2.1 REPRESENTING THE ELECTROMAGNETIC FIELD WITH DIFFERENTIAL FORMS

From Maxwell's laws in integral form, we can readily determine the degrees of the differential forms that will represent the various field quantities. In vector notation,

$$\oint_P \mathbf{E} \cdot d\mathbf{l} = -\frac{d}{dt}\int_A \mathbf{B} \cdot d\mathbf{A}$$

$$\oint_P \mathbf{H} \cdot d\mathbf{l} = \frac{d}{dt}\int_A \mathbf{D} \cdot d\mathbf{A} + \int_A \mathbf{J} \cdot d\mathbf{A}$$

$$\oint_S \mathbf{D} \cdot d\mathbf{S} = \int_V q\,dv$$

$$\oint_S \mathbf{B} \cdot d\mathbf{S} = 0$$

where A is a surface bounded by a path P, V is a volume bounded by a surface S, q is volume charge density, and the other quantities are defined as usual. The electric field intensity is integrated over a path, so that it becomes a 1-form. The magnetic field intensity is also integrated over a path, and becomes a 1-form as well. The electric

---

TABLE 7.1
Differential Forms of Each Degree

Degree	Region of Integration	Example	General Form
0-form	Point	$3x$	$f(x,y,z,\ldots)$
1-form	Path	$y^2\,dx + z\,dy$	$\alpha_1\,dx + \alpha_2\,dy + \alpha_3\,dz$
2-form	Surface	$e^y\,dy\,dz + e^x g\,dz\,dx$	$\beta_1\,dy\,dz + \beta_2\,dz\,dx + \beta_3\,dx\,dy$
3-form	Volume	$(x+y)\,dx\,dy\,dz$	$g\,dx\,dy\,dz$

TABLE 7.2

The Differential Forms that Represent Fields and Sources

Quantity	Form	Degree	Units	Vector/Scalar
Electric Field Intensity	E	1-form	V	**E**
Magnetic Field Intensity	H	1-form	A	**H**
Electric Flux Density	D	2-form	C	**D**
Magnetic Flux Density	B	2-form	Wb	**B**
Electric Current Density	J	2-form	A	**J**
Electric Charge Density	ρ	3-form	C	q

and magnetic flux densities are integrated over surfaces, and so are 2-forms. The sources are electric current density, which is a 2-form, since it falls under a surface integral, and the volume charge density, which is a 3-form, as it is integrated over a volume. Table 7.2 summarizes these forms.

7.2.2   1-Forms; Field Intensity

The usual physical motivation for electric field intensity is the force experienced by a small test charge placed in the field. This leads naturally to the vector representation of the electric field, which might be called the "force picture." Another physical viewpoint for the electric field is the change in potential experienced by a charge as it moves through the field. This leads naturally to the equipotential representation of the field, or the "energy picture." The energy picture shifts emphasis from the local concept of force experienced by a test charge to the global behavior of the field as manifested by change in energy of a test charge as it moves along a path.

Differential forms lead to the "energy picture" of field intensity. A 1-form is represented graphically as surfaces in space [2,3]. For a conservative field, the surfaces of the associated 1-form are equipotentials. The differential dx produces surfaces perpendicular to the x-axis, as shown in Figure 7.1a. Likewise, dy has surfaces perpendicular to the y-axis and the surfaces of dz are perpendicular to the z axis. A linear combination of these differentials has surfaces that are skew to the coordinate axes. The coefficients of a 1-form determine the spacing of the surfaces per unit length; the greater the magnitude of the coefficients, the more closely spaced are the surfaces. The 1-form $2dz$, shown in Figure 7.1b, has surfaces spaced twice as closely as those of dx in Figure 7.1a.

The surfaces of more general 1-forms can curve, end, or meet each other, depending on the behavior of the coefficients of the form. If surfaces of a 1-form do not meet or end, the field represented by the form is conservative. The field corresponding to the 1-form in Figure 7.1a is conservative; the field in Figure 7.1c is nonconservative.

Just as a line representing the magnitude of a vector has two possible orientations, the surfaces of a 1-form are oriented as well. This is done by specifying one of the two normal directions to the surfaces of the form. The surfaces of $3\,dx$ are oriented in the $+x$ direction, and those of $-3\,dx$ in the $-x$ direction. The orientation of a form is usually clear from context and is omitted from figures.

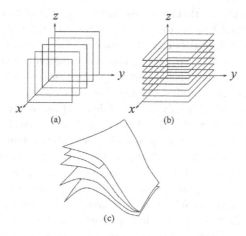

FIGURE 7.1   (a) The 1-form dx, with surfaces perpendicular to the x axis and infinite in the y and z directions. (b) The 1-form $2dz$, with surfaces perpendicular to the z-axis and spaced two per unit distance in the z direction. (c) A more general 1-form, with curved surfaces and surfaces that end or meet each other.

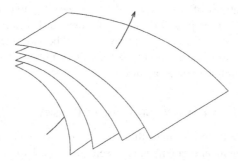

FIGURE 7.2   A path piercing four surfaces of a 1-form. The integral of the 1-form over the path is four.

Differential forms are by definition the quantities that can be integrated, so it is natural that the surfaces of a 1-form are a graphical representation of path integration. The integral of a 1-form along a path is the number of surfaces pierced by the path (Figure 7.2), taking into account the relative orientations of the surfaces and the path. This simple picture of path integration will provide in the next section a means for visualizing Ampere's and Faraday's laws.

The 1-form $E_1\,dx + E_2\,dy + E_3\,dz$ is said to be *dual* to the vector field $E_1\,\hat{\mathbf{x}} + E_2\,\hat{\mathbf{y}} + E_3\,\hat{\mathbf{z}}$. The field intensity 1-forms E and H are dual to the vectors $\mathbf{E}$ and $\mathbf{H}$.

Following Deschamps, we take the units of the electric and magnetic field intensity 1-forms to be Volts and Amps, as shown in Table 7.2. The differentials are considered to have units of length. Other field and source quantities are assigned units according to this same convention. A disadvantage of Deschamps' system is that it implies in a sense that the metric of space carries units. Alternative conventions are available; Bamberg and Sternberg [6] and others take the units of the electric and magnetic field intensity 1-forms to be V/m and A/m, the same as their vector counterparts, so that the differentials carry no units and the integration process itself is considered to provide a factor of length. If this convention is chosen, the basis differentials of curvilinear coordinate systems (see Section 7.4) must also be taken to carry no units. This leads to confusion for students, since these basis differentials can include factors of distance. The advantages of this alternative convention are that it is more consistent with the mathematical point of view, in which basis vectors and forms are abstract objects not associated with a particular system of units, and that a field quantity has the same units whether represented by a vector or a differential form. Furthermore, a general differential form may include differentials of functions that do not represent position and so cannot be assigned units of length. The possibility of confusion when using curvilinear coordinates seems to outweigh these considerations, and so we have chosen Deschamps' convention.

With this convention, the electric field intensity 1-form can be taken to have units of energy per charge, or J/C. This supports the "energy picture," in which the electric field represents the change in energy experienced by a charge as it moves through the field. One might argue that this motivation of field intensity is less intuitive than the concept of force experienced by a test charge at a point. While this may be true, the graphical representations of Ampere's and Faraday's laws that will be outlined in Section 7.3 favor the differential form point of view. Furthermore, the simple correspondence between vectors and forms allows both to be introduced with little additional effort, providing students a more solid understanding of field intensity than they could obtain from one representation alone.

7.2.3 2-Forms; Flux Density and Current Density

Flux density or flow of current can be thought of as tubes that connect sources of flux or current. This is the natural graphical representation of a 2-form, which is drawn as sets of surfaces that intersect to form tubes. The differential $dx\,dy$ is represented by the surfaces of dx and dy superimposed. The surfaces of dx perpendicular to the x-axis and those of dy perpendicular to the y-axis intersect to produce tubes in the z direction, as illustrated by Figure 7.3a. (To be precise, the tubes of a 2-form have no definite shape: tubes of $dx\,dy$ have the same density those of $[.5\,dx][2\,dy]$.) The coefficients of a 2-form give the spacing of the tubes. The greater the coefficients, the more dense the tubes. An arbitrary 2-form has tubes that may curve or converge at a point.

The direction of flow or flux along the tubes of a 2-form is given by the right-hand rule applied to the orientations of the surfaces making up the walls of a tube. The orientation of dx is in the $+x$ direction, and dy in the $+y$ direction, so the flux due to $dx\,dy$ is in the $+z$ direction.

As with 1-forms, the graphical representation of a 2-form is fundamentally related to the integration process. The integral of a 2-form over a surface is the number of tubes passing through the surface, where each tube is weighted positively if its

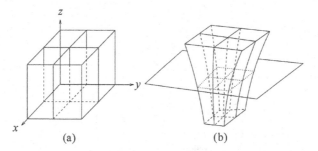

(a)   (b)

FIGURE 7.3 (a) The 2-form $dx\,dy$, with tubes in the z direction. (b) Four tubes of a 2-form pass through a surface, so that the integral of the 2-form over the surface is four.

orientation is in the direction of the surface's orientation, and negatively if opposite. This is illustrated in Figure 7.3b.

As with 1-forms, 2-forms correspond to vector fields in a simple way. An arbitrary 2-form $D_1\,dy\,dz + D_2\,dz\,dx + D_3\,dx\,dy$ is dual to the vector field $D_1\,\hat{x} + D_2\,\hat{y} + D_3\,\hat{z}$, so that the flux density 2-forms D and B are dual to the usual flux density vectors **D** and **B**.

7.2.4   3-Forms; Charge Density

Some scalar physical quantities are densities, and can be integrated over a volume. For other scalar quantities, such as electric potential, a volume integral makes no sense. The calculus of forms distinguishes between these two types of quantities by representing densities as 3-forms. Volume charge density, for example, becomes

$$\rho = q\,dx\,dy\,dz \qquad (7.1)$$

where q is the usual scalar charge density in the notation of [4].

A 3-form is represented by three sets of surfaces in space that intersect to form boxes. The density of the boxes is proportional to the coefficient of the 3-form; the greater the coefficient, the smaller and more closely spaced are the boxes. A point charge is represented by an infinitesimal box at the location of the charge. The 3-form $dx\,dy\,dz$ is the union of three families of planes perpendicular to each of the x, y and z axes. The planes along each of the axes are spaced one unit apart, forming cubes of unit side distributed evenly throughout space, as in Figure 7.4. The orientation of a 3-form is given by specifying the sign of its boxes. As with other differential forms, the orientation is usually clear from context and is omitted from figures.

7.2.5   0-Forms; Scalar Potential

0-forms are functions. The scalar potential ϕ, for example, is a 0-form. Any scalar physical quantity that is not a volume density is represented by a 0-form.

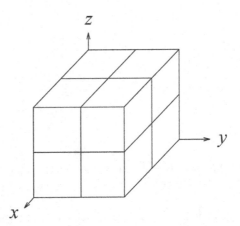

FIGURE 7.4   The 3-form $dx\, dy\, dz$, with cubes of side equal to one. The cubes fill all space.

7.2.6 SUMMARY

The use of differential forms helps students to understand electromagnetics by giving them distinct mental pictures that they can associate with the various fields and sources. As vectors, field intensity and flux density are mathematically and graphically indistinguishable as far as the type of physical quantity they represent. As differential forms, the two types of quantities have graphical representations that clearly express the physical meaning of the field. The surfaces of a field intensity 1-form assign potential change to a path. The tubes of a flux density 2-form give the total flux or flow through a surface. Charge density is also distinguished from other types of scalar quantities by its representation as a 3-form.

7.3   MAXWELL'S LAWS IN INTEGRAL FORM

In this section, we discuss Maxwell's laws in integral form in light of the graphical representations given in the previous section. Using the differential forms defined in Table 7.2, Maxwell's laws can be written

$$\oint_P E = -\frac{d}{dt}\int_A B$$
$$\oint_P H = \frac{d}{dt}\int_A D + \int_A J$$
$$\oint_S D = \int_V \rho \tag{7.2}$$
$$\oint_S B = 0$$

The first pair of laws is often more difficult for students to grasp than the second, because the vector picture of curl is not as intuitive as that for divergence. With differential forms, Ampere's and Faraday's laws are graphically very similar to Gauss's laws for the electric and magnetic fields. The close relationship between the two sets of laws becomes clearer.

7.3.1 AMPERE'S AND FARADAY'S LAWS

Faraday's and Ampere's laws equate the number of surfaces of a 1-form pierced by a closed path to the number of tubes of a 2-form passing through the path. Each tube of J, for example, must have a surface of H extending away from it, so that any path around the tube pierces the surface of H. Thus, Ampere's law states that tubes of displacement current and electric current are sources for surfaces of H. This is illustrated in Figure 7.5. Likewise, tubes of time-varying magnetic flux density are sources for surfaces of E.

The illustration of Ampere's law in Figure 7.5a is arguably the most important pedagogical advantage of the calculus of differential forms over vector analysis. Ampere's and Faraday's laws are usually considered the more difficult pair of Maxwell's laws, because vector analysis provides no simple picture that makes the physical meaning of these laws intuitive. Compare Figure 7.5a to the vector representation of the same field in Figure 7.5b. The vector field appears to "curl" everywhere in space. Students must be convinced that indeed the field has no curl except at the location of the current, using some pedagogical device such as an imaginary paddle wheel in a rotating fluid. The surfaces of H, on the other hand, end only along the tubes of current; where they do not end, the field has no curl. This is the fundamental concept underlying Ampere's and Faraday's laws: tubes of time-varying flux or current produce field intensity surfaces.

7.3.2 GAUSS'S LAWS

Gauss's law for the electric field states that the number of tubes of D flowing out through a closed surface must be equal to the number of boxes of ρ inside the surface.

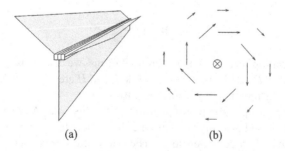

(a)   (b)

FIGURE 7.5 (a) A graphical representation of Ampere's law: tubes of current produce surfaces of magnetic field intensity. Any loop around the three tubes of J must pierce three surfaces of H. (b) A cross-section of the same magnetic field using vectors. The vector field appears to "curl" everywhere, even though the field has nonzero curl only at the location of the current.

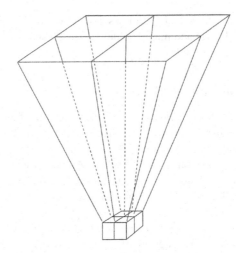

FIGURE 7.6 A graphical representation of Gauss's law for the electric flux density: boxes of ρ produce tubes of D.

The boxes of ρ are sources for the tubes of D, as shown in Figure 7.6. Gauss's law for the magnetic flux density states that tubes of the 2-form B can never end—they must either form closed loops or go off to infinity.

Comparing Figures. 7.5a and 7.6 shows the close relationship between the two sets of Maxwell's laws. In the same way that flux density tubes are produced by boxes of electric charge, field intensity surfaces are produced by tubes of the sources on the right-hand sides of Faraday's and Ampere's laws. Conceptually, the laws only differ in the degrees of the forms involved and the dimensions of their pictures.

7.3.3 Constitutive Relations and the Star Operator

The vector expressions of the constitutive relations for an isotropic medium,

$$\mathbf{D} = \epsilon\,\mathbf{E}$$

$$\mathbf{B} = \mu\mathbf{H}$$

involve scalar multiplication. With differential forms, we cannot use these same relationships, because D and B are 2-forms, while E and H are 1-forms. An operator that relates forms of different degrees must be introduced.

The Hodge star operator [6,18] naturally fills this role. As vector spaces, the spaces of 0-forms and 3-forms are both one-dimensional, and the spaces of 1-forms and 2-forms are both three-dimensional. The star operator $\star$ is a set of isomorphisms between these pairs of vector spaces.

For 1-forms and 2-forms, the star operator satisfies

$$\star dx = dy\,dz$$

$$\star dy = dz\, dx$$

$$\star dz = dx\, dy$$

0-forms and 3-forms are related by

$$\star 1 = dx\, dy\, dz$$

In R^3, the star operator is its own inverse, so that $\star\star\alpha = \alpha$. A 1-form ω is dual to the same vector as the 2-form $\star\omega$.

Graphically, the star operator replaces the surfaces of a form with orthogonal surfaces, as in Figure 7.7. The 1-form $3\,dx$, for example, has planes perpendicular to the x-axis. It becomes $3\,dy\,dz$ under the star operation. This 2-form has planes perpendicular to the y and the z axes.

By using the star operator, the constitutive relations can be written as

$$D = \epsilon \star E \tag{7.3}$$

$$B = \mu \star H \tag{7.4}$$

where ϵ and μ are the permittivity and permeability of the medium. The surfaces of E are perpendicular to the tubes of D, and the surfaces of H are perpendicular to the tubes of B. The following example illustrates the use of these relations.

7.3.4 THE EXTERIOR PRODUCT AND THE POYNTING 2-FORM

Between the differentials of 2-forms and 3-forms is an implied exterior product, denoted by a wedge $\wedge$. The wedge is nearly always omitted from the differentials of a form, especially when the form appears under an integral sign. The exterior product of 1-forms is anticommutative, so that $dx \wedge dy = -dy \wedge dx$. As a consequence, the exterior product is in general supercommutative:

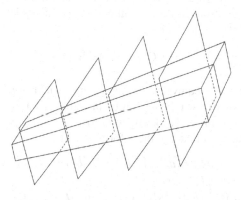

FIGURE 7.7 The star operator relates 1-form surfaces to perpendicular 2-form tubes.

Example 7.1: Finding D due to an electric field intensity

Let $E = (dx + dy)e^{ik(x-y)}$ V be the electric field in free space. We wish to find the flux density due to this field. Using the constitutive relationship between D and E,

$$D = \epsilon_0 \star \left(dx + dy \right) e^{ik(x-y)} = \epsilon_0 e^{ik(x-y)} \left(\star dx + \star dy \right) = \epsilon_0 e^{ik(x-y)} \left(dy\,dz + dz\,dx \right) C$$

While we restrict our attention to isotropic media, the star operator applies equally well to anisotropic media. As discussed in Ref. [6] and elsewhere, the star operator depends on a metric. If the metric is related to the permittivity or the permeability tensor in an appropriate manner, anisotropic star operators are obtained, and the constitutive relations become $D = \star_e E$ and $B = \star_h H$. Graphically, an anisotropic star operator acts on 1-form surfaces to produce 2-form tubes that intersect the surfaces obliquely rather than orthogonally.

$$\alpha \wedge \beta = \left(-1\right)^{ab} \beta \wedge \alpha \tag{7.5}$$

where a and b are the degrees of α and β, respectively. One usually converts the differentials of a form to right–cyclic order using (7.5).

As a consequence of (7.5), any differential form with a repeated differential vanishes. In a three-dimensional space each term of a p-form will always contain a repeated differential if $p > 3$, so there are no nonzero p-forms for $p > 3$.

The exterior product of two 1-forms is analogous to the vector cross product. With vector analysis, it is not obvious that the cross product of vectors is a different type of quantity than the factors. Under coordinate inversion, $\mathbf{A} \times \mathbf{B}$ changes sign relative to a vector with the same components, so that $\mathbf{A} \times \mathbf{B}$ is a pseudovector. With forms, the distinction between $a \wedge b$ and a or b individually is clear.

The exterior product of a 1-form and a 2-form corresponds to the dot product. The coefficient of the resulting 3-form is equal to the dot product of the vector fields dual to the 1-form and 2-form in the euclidean metric.

Combinations of cross and dot products are somewhat difficult to manipulate algebraically, often requiring the use of tabulated identities. Using the supercommutativity of the exterior product, the student can easily manipulate arbitrary products of forms. For example, the identities

$$\mathbf{A} \cdot \left(\mathbf{B} \times \mathbf{C} \right) = \mathbf{C} \cdot \left(\mathbf{A} \times \mathbf{B} \right) = \mathbf{B} \cdot \left(\mathbf{C} \times \mathbf{A} \right)$$

are in the euclidean metric equivalent to relationships which are easily obtained from (7.5). Factors in any exterior product can be interchanged arbitrarily as long as the sign of the product is changed according to (7.5).

Consider the exterior product of the 1-forms E and ,

$$\begin{aligned}
E \wedge H &= \left(E_1 dx + E_2 dy + E_3 dz \right) \wedge \left(H_1 dx + H_2 dy + H_3 dz \right) \\
&= E_1 H_1 dx\,dx + E_1 H_2 dx\,dy + E_1 H_3 dx\,dz + E_2 H_1 dy\,dx + E_2 H_2 dy\,dy \\
&\quad + E_2 H_3 dy\,dz + E_3 H_1 dz\,dx + E_3 H_2 dz\,dy + E_3 H_3 dz\,dz \\
&= \left(E_2 H_3 - E_3 H_2 \right) dy\,dz + \left(E_3 H_1 - E_1 H_3 \right) dz\,dx + \left(E_1 H_2 - E_3 H_1 \right) dx\,dy
\end{aligned}$$

Power

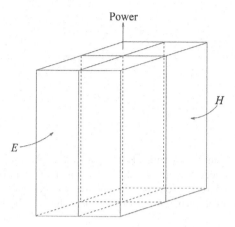

FIGURE 7.8   The Poynting power flow 2-form $S = E \wedge H$. Surfaces of the 1-forms E and H are the sides of the tubes of S.

This is the Poynting 2-form S. For complex fields, $S = E \wedge H^*$. For time-varying fields, the tubes of this 2-form represent flow of electromagnetic power, as shown in Figure 7.8. The sides of the tubes are the surfaces of E and H. This gives a clear geometrical interpretation to the fact that the direction of power flow is orthogonal to the orientations of both E and H.

Example 7.2:   The Poynting 2-form due to a plane wave

Consider a plane wave propagating in free space in the z direction, with the time–harmonic electric field $E = E_0 e^{jkz}\, dx$ V in the x direction. The complex Poynting 2-form is

$$
\begin{aligned}
S &= E \wedge H^* \\
&= E_0\, dx \wedge \frac{E_0}{\eta_0}\, dy \\
&= \frac{|E_0|^2}{\eta_0}\, dx\, dy\ (W)
\end{aligned}
$$

where η_0 is the wave impedance of free space.

7.3.5   ENERGY DENSITY

The exterior products $E \wedge D$ and $H \wedge B$ are 3-forms that represent the density of electromagnetic energy. The energy density 3-form w is defined to be

$$
w = \frac{1}{2}\left(E \wedge D + H \wedge B\right)
\tag{7.6}
$$

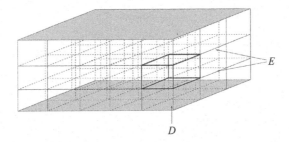

FIGURE 7.9 The 3-form $2w$ due to fields inside a parallel plate capacitor with oppositely charged plates. The surfaces of E are parallel to the top and bottom plates. The tubes of D extend vertically from charges on one plate to opposite charges on the other. The tubes and surfaces intersect to form cubes of 2ω, one of which is outlined in the figure.

The volume integral of w gives the total energy stored in a region of space by the fields present in the region.

Figure 7.9 shows the energy density 3-form between the plates of a capacitor, where the upper and lower plates are equally and oppositely charged. The boxes of $2w$ are the intersection of the surfaces of E, which are parallel to the plates, with the tubes of D, which extend vertically from one plate to the other.

7.4 CURVILINEAR COORDINATE SYSTEMS

In this section, we give the basis differentials, the star operator, and the correspondence between vectors and forms for cylindrical, spherical, and generalized orthogonal coordinates.

7.4.1 CYLINDRICAL COORDINATES

The differentials of the cylindrical coordinate system are $d\rho$, $\rho\,d\phi$ and dz. Each of the basis differentials is considered to have units of length. The general 1-form

$$A\,d\rho + B\rho\,d\phi + C\,dz \qquad (7.7)$$

is dual to the vector

$$A\hat{\rho} + B\hat{\phi} + C\hat{z} \qquad (7.8)$$

The general 2-form

$$A\rho\,d\phi \wedge dz + B\,dz \wedge d\rho + C\,d\rho \wedge \rho\,d\phi \qquad (7.9)$$

is dual to the same vector. The 2-form $d\rho\,d\phi$, for example, is dual to the vector $(1/\rho)z$.

Differentials must be converted to basis elements before the star operator is applied. The star operator in cylindrical coordinates acts as follows:

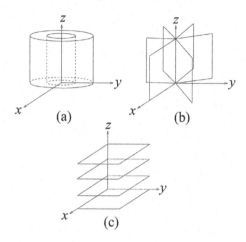

FIGURE 7.10 Surfaces of (a) $d\rho$, (b) $d\phi$ scaled by $3/\pi$, and (c) dz.

$$\star d\rho = \rho \, d\phi \wedge dz$$

$$\star \rho \, d\phi = dz \wedge d\rho$$

$$\star dz = d\rho \wedge \rho \, d\phi$$

Also, $\star 1 = \rho \, d\rho \, d\phi \, dz$. As with the rectangular coordinate system, $\star \star = 1$. The star operator applied to $d\phi \, dz$, for example, yields $(1/\rho) \, d\rho$.

Figure 7.10 shows the pictures of the differentials of the cylindrical coordinate system. The 2-forms can be obtained by superimposing these surfaces. Tubes of $dz \wedge d\rho$, for example, are square rings formed by the union of Figures. 7.10a and 7.10c.

7.4.2 Spherical Coordinates

The basis differentials of the spherical coordinate system are in right-cyclic order, dr, $r \, d\theta$ and $r \sin \theta \, d\phi$, each having units of length. The 1-form

$$A \, dr + B r \, d\theta + C r \sin \theta \, d\phi \tag{7.10}$$

and the 2-form

$$A r \, d\theta \wedge r \sin \theta \, d\phi + B r \sin \theta \, d\phi \wedge dr + C \, dr \wedge r \, d\theta \tag{7.11}$$

are both dual to the vector

$$A \hat{\mathbf{r}} + B \hat{\boldsymbol{\theta}} + C \hat{\boldsymbol{\phi}} \tag{7.12}$$

so that $d\theta \, d\phi$, for example, is dual to the vector $r/(r^2 \sin \theta)$.

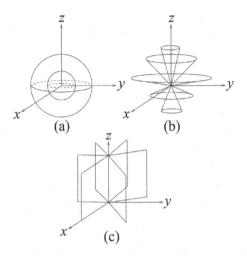

FIGURE 7.11   Surfaces of (a) dr, (b) $d\theta$ scaled by $10/\pi$, and (c) $d\phi$ scaled by $3/\pi$.

As in the cylindrical coordinate system, differentials must be converted to basis elements before the star operator is applied. The star operator acts on 1-forms and 2-forms as follows:

$$\star dr = r\, d\theta \wedge r\sin\theta\, d\phi$$

$$\star r\, d\theta = r\sin\theta\, d\phi \wedge dr$$

$$\star r\sin\theta\, d\phi = dr \wedge r\, d\theta$$

Again, $\star\star = 1$. The star operator applied to one is $\star 1 = r^2 \sin\theta\, dr\, d\theta\, d\phi$. Figure 7.11 shows the pictures of the differentials of the spherical coordinate system; pictures of 2-forms can be obtained by superimposing these surfaces.

7.4.3   GENERALIZED ORTHOGONAL COORDINATES

Let the location of a point be given by (u, v, w) such that the tangents to each of the coordinates are mutually orthogonal. Define a function h_1 such that the integral of $h_1\, du$ along any path with v and w constant gives the length of the path. Define h_2 and h_3 similarly. Then the basis differentials are

$$h_1\, du, h_2\, dv, h_3\, dw \tag{7.13}$$

The 1-form $Ah_1\, du + Bh_2\, dv + Ch_3\, dw$ and the 2-form $Ah_2h_3\, dv \wedge dw + Bh_3h_1\, dw \wedge du + Ch_1h_2\, du \wedge dv$ are both dual to the vector $A\hat{\mathbf{u}} + B\hat{\mathbf{v}} + C\hat{\mathbf{w}}$. The star operator on 1-forms and 2-forms satisfies

$$\star\left(Ah_1 du + Bh_2 dv + Ch_3 dw\right) = Ah_2h_3 dv \wedge dw + Bh_3h_1 dw \wedge du + Ch_1h_2 du \wedge dv \tag{7.14}$$

For 0-forms and 3-forms, $\star 1 = h_1h_2h_3\, du\, dv\, dw$.

7.5   ELECTROSTATICS AND MAGNETOSTATICS

In this section we treat several of the usual elementary applications of Maxwell's laws in integral form. We find the electric flux due to a point charge and a line charge using Gauss's law for the electric field. Ampere's law is used to find the magnetic fields produced by a line current.

7.5.1   Point Charge

By symmetry, the tubes of flux from a point charge Q must extend out radially from the charge (Figure 7.12), so that

$$D = D_0 r^2 \sin\theta \, d\theta \, d\phi \tag{7.15}$$

To apply Gauss law $\oint_S D = \int_V \rho$, we choose S to be a sphere enclosing the charge. The right-hand side of Gauss's law is equal to Q, and the left-hand side is

$$
\begin{aligned}
\oint_S D &= \int_0^{2\pi}\int_0^{\pi} D_0 \; r^2 \; \sin\theta \; d\theta \; d\phi \\
&= 4\pi r^2 D_0
\end{aligned}
$$

Solving for D_0 and substituting into (7.15),

$$D = \frac{Q}{4\pi r^2} r \, d\theta \, r \sin\theta \, d\phi \, C \tag{7.16}$$

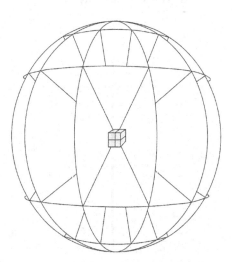

FIGURE 7.12   Electric flux density due to a point charge. Tubes of D extend away from the charge.

for the electric flux density due to the point charge. This can also be written

$$D = \frac{Q}{4\pi} \sin\theta \, d\theta \, d\phi \ C \qquad (7.17)$$

Since 4π is the total amount of solid angle for a sphere and $\sin\theta \, d\theta \, d\phi$ is the differential element of solid angle, this expression matches Figure 7.12 in showing that the amount of flux per solid angle is constant.

7.5.2 LINE CHARGE

For a line charge with charge density ρ_l C/m, by symmetry tubes of flux extend out radially from the line, as shown in Figure 7.13. The tubes are bounded by the surfaces of $d\phi$ and dz, so that D has the form

$$D = D_0 \, d\phi \, dz \qquad (7.18)$$

Let S be a cylinder of height b with the line charge along its axis. The right-hand side of Gauss's law is

$$\int_V \rho \ = \ \int_0^b \rho_l \ dz$$
$$= \ b\rho_l$$

The left-hand side is

$$\oint_S D = \int_0^b \int_0^{2\pi} D_0 \ d\phi \ dz$$
$$= 2\pi b \ D_0$$

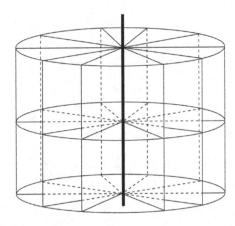

FIGURE 7.13 Electric flux density due to a line charge. Tubes of D extend radially away from the vertical line of charge.

Solving for D_0 and substituting into (7.18), we obtain

$$D = \frac{\rho_l}{2\pi} d\phi \, dz \, C \tag{7.19}$$

for the electric flux density due to the line charge.

7.5.3 LINE CURRENT

If a current I_l A flows along the z-axis, sheets of the H 1-form will extend out radially from the current, as shown in Figure 7.14. These are the surfaces of $d\phi$, so that by symmetry,

$$H = H_0 \, d\phi \tag{7.20}$$

where H_0 is a constant we need to find using Ampere's law. We choose the path P in Ampere's law $\oint_P H = \frac{d}{dt} \int_A D + \int_A J$ to be a loop around the z-axis. Assuming that $D = 0$, the right-hand side of Ampere's law is equal to I_l. The left-hand side is the integral of H over the loop,

$$\oint_P H = \int_0^{2\pi} H_0 \, d\phi$$

$$= 2\pi \, H_0$$

The magnetic field intensity is then

$$H = \frac{I_l}{2\pi} d\phi \, A \tag{7.21}$$

for the line current source.

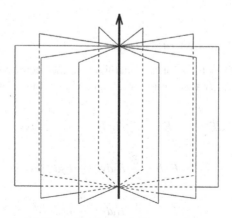

FIGURE 7.14 Magnetic field intensity H due to a line current.

7.6 THE EXTERIOR DERIVATIVE AND MAXWELL'S LAWS IN POINT FORM

In this section we introduce the exterior derivative and the generalized Stokes theorem and use these to express Maxwell's laws in point form. The exterior derivative is a single operator which has the gradient, curl, and divergence as special cases, depending on the degree of the differential form on which the exterior derivative acts. The exterior derivative has the symbol d, and can be written formally as

$$d \equiv \frac{\partial}{\partial x} dx + \frac{\partial}{\partial y} dy + \frac{\partial}{\partial z} dz \tag{7.22}$$

The exterior derivative can be thought of as implicit differentiation with new differentials introduced from the left.

7.6.1 EXTERIOR DERIVATIVE OF 0-FORMS

Consider the 0-form $f(x, y, z)$. If we implicitly differentiate f with respect to each of the coordinates, we obtain

$$df = \frac{\partial f}{\partial x} dx + \frac{\partial f}{\partial y} dy + \frac{\partial f}{\partial z} dz \tag{7.23}$$

which is a 1-form, the exterior derivative of f. Note that the differentials dx, dy, and dz are the exterior derivatives of the coordinate functions x, y, and z. The 1-form df is dual to the gradient of f.

If ϕ represents a scalar electric potential, the negative of its exterior derivative is electric field intensity:

$$E = -d\phi$$

As noted earlier, the surfaces of the 1-form E are equipotentials, or level sets of the function ϕ, so that the exterior derivative of a 0-form has a simple graphical interpretation.

7.6.2 EXTERIOR DERIVATIVE OF 1-FORMS

The exterior derivative of a 1-form is analogous to the vector curl operation. If E is an arbitrary 1-form $E_1 \, dx + E_2 \, dy + E_3 \, dz$, then the exterior derivative of E is

$$dE = \left(\frac{\partial}{\partial x} E_1 dx + \frac{\partial}{\partial y} E_1 dy + \frac{\partial}{\partial z} E_1 dz \right) dx$$

$$= \left(\frac{\partial}{\partial x} E_2 dx + \frac{\partial}{\partial y} E_2 dy + \frac{\partial}{\partial z} E_2 dz \right) dy$$

$$= \left(\frac{\partial}{\partial x} E_3 dx + \frac{\partial}{\partial y} E_3 dy + \frac{\partial}{\partial z} E_3 dz \right) dz$$

Using the antisymmetry of the exterior product, this becomes

$$dE = \left(\frac{\partial E_3}{\partial y} - \frac{\partial E_2}{\partial z}\right) dy\,dz + \left(\frac{\partial E_1}{\partial z} - \frac{\partial E_3}{\partial x}\right) dz\,dx + \left(\frac{\partial E_2}{\partial x} - \frac{\partial E_1}{\partial y}\right) dx\,dy \qquad (7.24)$$

which is a 2-form dual to the curl of the vector field $E_1\hat{x} + E_2\hat{y} + E_3\hat{z}$.

Any 1-form E for which $dE = 0$ is called *closed* and represents a conservative field. Surfaces representing different potential values can never meet. If $dE \neq 0$, the field is nonconservative, and surfaces meet or end wherever the exterior derivative is nonzero.

7.6.3 EXTERIOR DERIVATIVE OF 2-FORMS

The exterior derivative of a 2-form is computed by the same rule as for 0-forms and 1-forms: take partial derivatives by each coordinate variable and add the corresponding differential on the left. For an arbitrary 2-form B,

$$\begin{aligned}
dB &= d\left(B_1\,dy\,dz + B_2\,dz\,dx + B_3\,dx\,dy\right) \\
&= \left(\frac{\partial}{\partial x} B_1\,dx + \frac{\partial}{\partial y} B_1\,dy + \frac{\partial}{\partial z} B_1\,dz\right) dy\,dz \\
&\quad + \left(\frac{\partial}{\partial x} B_2\,dx + \frac{\partial}{\partial y} B_2\,dy + \frac{\partial}{\partial z} B_2\,dz\right) dz\,dx \\
&\quad + \left(\frac{\partial}{\partial x} B_3\,dx + \frac{\partial}{\partial y} B_3\,dy + \frac{\partial}{\partial z} B_3\,dz\right) dx\,dy \\
&= \left(\frac{\partial B_1}{\partial x} + \frac{\partial B_2}{\partial y} + \frac{\partial B_3}{\partial z}\right) dx\,dy\,dz
\end{aligned}$$

where six of the terms vanish due to repeated differentials. The coefficient of the resulting 3-form is the divergence of the vector field dual to B.

7.6.4 PROPERTIES OF THE EXTERIOR DERIVATIVE

Because the exterior derivative unifies the gradient, curl, and divergence operators, many common vector identities become special cases of simple properties of the exterior derivative. The equality of mixed partial derivatives leads to the identity

$$dd = 0 \qquad (7.25)$$

so that the exterior derivative applied twice yields zero. This relationship is equivalent to the vector relationships $\nabla \times (\nabla f) = 0$ and $\nabla \cdot (\nabla \times \mathbf{A}) = 0$. The exterior derivative also obeys the product rule

$$d(\alpha \wedge \beta) = d\alpha \wedge \beta + (-1)^p \alpha \wedge d\beta \qquad (7.26)$$

where p is the degree of α. A special case of (7.26) is

$$\nabla \cdot (\mathbf{A} \times \mathbf{B}) = \mathbf{B} \cdot (\nabla \times \mathbf{A}) - \mathbf{A} \cdot (\nabla \times \mathbf{B})$$

These and other vector identities are often placed in reference tables; by contrast, (7.25) and (7.26) are easily remembered.

The exterior derivative in cylindrical coordinates is

$$d = \frac{\partial}{\partial \rho} d\rho + \frac{\partial}{\partial \phi} d\phi + \frac{\partial}{\partial z} dz \tag{7.27}$$

which is the same as for rectangular coordinates but with the coordinates ρ, ϕ, z in the place of x, y, z. Note that the exterior derivative does not require the factor of ρ that is involved in converting forms to vectors and applying the star operator. In spherical coordinates,

$$d = \frac{\partial}{\partial r} dr + \frac{\partial}{\partial \theta} d\theta + \frac{\partial}{\partial \phi} d\phi \tag{7.28}$$

where the factors r and $r \sin \theta$ are not found in the exterior derivative operator. The exterior derivative is

$$d = \frac{\partial}{\partial u} du + \frac{\partial}{\partial v} dv + \frac{\partial}{\partial w} dw \tag{7.29}$$

in general orthogonal coordinates. The exterior derivative is much easier to apply in curvilinear coordinates than the vector derivatives; there is no need for reference tables of derivative formulas in various coordinate systems.

7.6.5 THE GENERALIZED STOKES THEOREM

The exterior derivative satisfies the generalized Stokes theorem, which states that for any p-form ω,

$$\int_M d\omega = \oint_{bd\, M} \omega \tag{7.30}$$

where M is a $(p + 1)$–dimensional region of space and $bd\, M$ is its boundary. If ω is a 0-form, then the Stokes theorem becomes $\int_a^b df = f(b) - f(a)$. This is the fundamental theorem of calculus.

If ω is a 1-form, then $bd\, M$ is a closed loop and M is a surface that has the path as its boundary. This case is analogous to the vector Stokes theorem. Graphically, the number of surfaces of ω pierced by the loop equals the number of tubes of the 2-form $d\omega$ that pass through the loop (Figure 7.15).

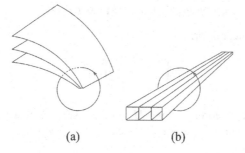

FIGURE 7.15   The Stokes theorem for ω a 1-form. (a) The loop $bd\,M$ pierces three of the surfaces of ω. (b) Three tubes of $d\omega$ pass through any surface M bounded by the loop $bd\,M$.

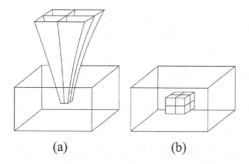

FIGURE 7.16   Stokes theorem for ω a 2-form. (a) Four tubes of the 2-form ω pass through a surface. (b) The same number of boxes of the 3-form $d\omega$ lie inside the surface.

If ω is a 2-form, then $bd\,M$ is a closed surface and M is the volume inside it. The Stokes theorem requires that the number of tubes of ω that cross the surface equal the number of boxes of $d\omega$ inside the surface, as shown in Figure 7.16. This is equivalent to the vector divergence theorem.

Compared to the usual formulations of these theorems,

$$f(b) - f(a) = \int_a^b \frac{\partial f}{\partial x}\,dx$$

$$\oint_{bd\,A} \mathbf{E} \cdot d\mathbf{l} = \int_A \nabla \times \mathbf{E} \cdot d\mathbf{A}$$

$$\oint_{bd\,V} \mathbf{D} \cdot d\mathbf{S} = \int_V \nabla \cdot \mathbf{D}\,dv$$

the generalized Stokes theorem is simpler in form and hence easier to remember. It also makes clear that the vector Stokes theorem and the divergence theorem are higher-dimensional statements of the fundamental theorem of calculus.

7.6.6 FARADAY'S AND AMPERE'S LAWS IN POINT FORM

Faraday's law in integral form is

$$\oint_P E = -\frac{d}{dt}\int_A B \tag{7.31}$$

Using the Stokes theorem, taking M to be the surface A, we can relate the path integral of E to the surface integral of the exterior derivative of E,

$$\oint_P E = \int_A dE \tag{7.32}$$

By Faraday's law,

$$\int_A dE = -\frac{d}{dt}\int_A B \tag{7.33}$$

For sufficiently regular forms E and B, we have that

$$dE = -\frac{\partial B}{\partial t} \tag{7.34}$$

since (7.33) is valid for all surfaces A. This is Faraday's law in point form. This law states that new surfaces of E are produced by tubes of time-varying magnetic flux.

Using the same argument, Ampere's law becomes

$$dH = \frac{\partial D}{\partial t} + J \tag{7.35}$$

Ampere's law shows that new surfaces of H are produced by tubes of time-varying electric flux or electric current.

7.6.7 GAUSS'S LAWS IN POINT FORM

Gauss's law for the electric flux density is

$$\oint_S D = \int_V \rho \tag{7.36}$$

The Stokes theorem with M as the volume V and $bd\, M$ as the surface S shows that

$$\oint_S D = \int_V dD \tag{7.37}$$

Using Gauss's law in integral form (7.36),

$$\int_V dD = \int_V \rho \tag{7.38}$$

We can then write

$$dD = \rho \tag{7.39}$$

This is Gauss's law for the electric field in point form. Graphically, this law shows that tubes of electric flux density can end only on electric charges. Similarly, Gauss's law for the magnetic field is

$$dB = 0 \tag{7.40}$$

This law requires that tubes of magnetic flux density never end; they must form closed loops or extend to infinity.

7.6.8 POYNTING'S THEOREM

Using Maxwell's laws, we can derive a conservation law for electromagnetic energy. The exterior derivative of S is

$$dS = d(E \wedge H)$$
$$= (dE) \wedge H - E \wedge (dH)$$

Using Ampere's and Faraday's laws, this can be written

$$dS = -\frac{\partial B}{\partial t} \wedge H - E \wedge \frac{\partial D}{\partial t} - E \wedge J \tag{7.41}$$

Finally, using the definition (7.6) of w, this becomes

$$dS = -\frac{\partial w}{\partial t} - E \wedge J \tag{7.42}$$

At a point where no sources exist, a change in stored electromagnetic energy must be accompanied by tubes of S that represent flow of energy toward or away from the point.

7.6.9 INTEGRATING FORMS BY PULLBACK

We have seen in previous sections that differential forms give integration a clear graphical interpretation. The use of differential forms also results in several simplifications of the integration process itself. Integrals of vector fields require a metric; integrals of differential forms do not. The method of pullback replaces the computation of differential length and surface elements that is required before a vector field can be integrated.

Consider the path integral

$$\int_P \mathbf{E} \cdot d\mathbf{l} \tag{7.43}$$

The dot product of $\mathbf{E}$ with $d\mathbf{l}$ produces a 1-form with a single differential in the parameter of the path P, allowing the integral to be evaluated. The integral of the 1-form E dual to $\mathbf{E}$ over the same path is computed by the method of *pullback*, as change of variables for differential forms is commonly termed. Let the path P be parameterized by

$$x = p_1(t), y = p_2(t), z = p_3(t)$$

for $a < t < b$. The pullback of E to the path P is denoted $P*E$, and is defined to be

$$\begin{aligned}
P * E &= P * \left(E_1 \, dx + E_2 \, dy + E_3 \, dz \right) \\
&= E_1 \left(p_1, p_2, p_3 \right) d p_1 + E_2 \left(p_1, p_2, p_3 \right) d p_2 + E_3 \left(p_1, p_2, p_3 \right) d p_3 \\
&= \left(E_1 \left(p_1, p_2, p_3 \right) \frac{\partial p_1}{\partial t} + E_2 \left(p_1, p_2, p_3 \right) \frac{\partial p_2}{\partial t} + E_3 \left(p_1, p_2, p_3 \right) \frac{\partial p_3}{\partial t} \right) dt
\end{aligned}$$

Using the pullback of E, we convert the integral over P to an integral in t over the interval $[a, b]$,

$$\int_P E = \int_a^b P^* E \tag{7.44}$$

Components of the Jacobian matrix of the coordinate transform from the original coordinate system to the parameterization of the region of integration enter naturally when the exterior derivatives are performed. Pullback works similarly for 2-forms and 3-forms, allowing evaluation of surface and volume integrals by the same method. The following example illustrates the use of pullback.

Example 7.3: Work required to move a charge through an electric field

Let the electric field intensity be given by $E = 2xy \, dx + x^2 \, dy - dz$. A charge of $q = 1$ C is transported over the path P given by $(x = t^2, y = t, z = 1 - t^3)$ from $t = 0$ to $t = 1$. The work required is given by

$$W = -q \int_P 2xy \, dx + x^2 \, dy - dz \tag{7.45}$$

which by Equation (7.44) is equal to

$$-q \int_0^1 P^* \left(2xy \, dx + x^2 \, dy - dz \right)$$

where $P*E$ is the pullback of the field 1-form to the path P,

$$P * E = 2\left(t^2\right)(t)\ 2t\ dt + \left(t^2\right)^2 dt - \left(-3t^2\right) dt$$
$$= \left(5t^4 + 3t^2\right) dt$$

Integrating this new 1-form in t over $[0, 1]$, we obtain

$$W = -\int_0^1 \left(5t^4 + 3t^2\right) dt = -2\ J$$

as the total work required to move the charge along P.

7.6.10 EXISTENCE OF GRAPHICAL REPRESENTATIONS

With the exterior derivative, a condition can be given for the existence of the graphical representations of Section 7.2. These representations do not correspond to the
usual "tangent space" picture of a vector field, but rather are analogous to the integral
curves of a vector field. Obtaining the graphical representation of a differential form
as a family of surfaces is in general nontrivial, and is closely related to Pfaff's problem [19]. By the solution to Pfaff's problem, each differential form may be represented graphically in two dimensions as families of lines. In three dimensions, a
1-form ω can be represented as surfaces if the rotation $\omega \wedge d\omega$ is zero. If $\omega \wedge d\omega \neq 0$,
then there exist local coordinates for which ω has the form $du + v\ dw$, so that it is the
sum of two 1-forms, both of which can be graphically represented as surfaces.

An arbitrary, smooth 2-form in $\mathbf{R}^3$ can be written locally in the form $fdg \wedge dh$, so
that the 2-form consists of tubes of $dg \wedge dh$ scaled by f.

7.6.11 SUMMARY

Throughout this section, we have noted various aspects of the calculus of differential
forms that simplify manipulations and provide insight into the principles of electromagnetics. The exterior derivative behaves differently depending on the degree of the
form it operates on, so that physical properties of a field are encoded in the type of
form used to represent it, rather than in the type of operator used to take its derivative.
The generalized Stokes theorem gives the vector Stokes theorem and the divergence
theorem intuitive graphical interpretations that illuminate the relationship between
the two theorems. While of lesser pedagogical importance, the algebraic and computational advantages of forms cited in this section also aid students by reducing the
need for reference tables or memorization of identities.

7.7 THE INTERIOR PRODUCT AND BOUNDARY CONDITIONS

Boundary conditions can be expressed using a combination of the exterior and interior products. The same operator is used to express boundary conditions for field
intensities and flux densities, and in both cases the boundary conditions have simple
graphical interpretations.

7.7.1 THE INTERIOR PRODUCT

The interior product has the symbol ⌟. Graphically, the interior product removes the surfaces of the first form from those of the second. The interior product $dx \lrcorner dy$ is zero (in the Euclidian metric), since there are no dx surfaces to remove. The interior product of dx with itself is one. The interior product of dx and $dx\, dy$ is $dx \lrcorner dx\, dy = dy$. To compute the interior product $dy \lrcorner dx\, dy$, the differential dy must be moved to the left of $dx\, dy$ before it can be removed, so that

$$dy \lrcorner dx\, dy = -dy \lrcorner dy\, dx = -dx$$

The interior product of arbitrary 1-forms can be found by linearity from the relationships

$$
\begin{aligned}
dx \ \lrcorner \ dx &= 1, \ dx \ \lrcorner \ dy = 0, \ dx \ \lrcorner \ dz = 0 \\
dy \ \lrcorner \ dx &= 0, \ dy \ \lrcorner \ dy = 1, \ dy \ \lrcorner \ dz = 0 \\
dz \ \lrcorner \ dx &= 0, \ dz \ \lrcorner \ dy = 0, \ dz \ \lrcorner \ dz = 1
\end{aligned}
\tag{7.46}
$$

The interior product of a 1-form and a 2-form can be found using

$$
\begin{aligned}
dx \ \lrcorner \ dy \wedge dz &= 0, \quad dx \ \lrcorner \ dz \wedge dx = -dz, \quad dx \ \lrcorner \ dx \wedge dy = dy \\
dy \ \lrcorner \ dy \wedge dz &= dz, \quad dy \ \lrcorner \ dz \wedge dx = 0, \quad dy \ \lrcorner \ dx \wedge dy = -dx \\
dz \ \lrcorner \ dy \wedge dz &= -dy, \quad dz \ \lrcorner \ dz \wedge dx = dx, \quad dz \ \lrcorner \ dx \wedge dy = 0
\end{aligned}
\tag{7.47}
$$

The following examples illustrate the use of the interior product.

Example 7.4: The interior product of two 1-forms

The interior product of $a = 3x\, dx - y\, dz$ and $b = 4\, dy + 5\, dz$ is

$$
\begin{aligned}
a \lrcorner b &= \left(3x\, dx - y\, dz\right) \lrcorner \left(4\, dy + 5\, dz\right) \\
&= 12x\, dx \lrcorner dy + 15x\, dx \lrcorner dz - 4y\, dz \lrcorner dy - 5y\, dz \lrcorner dz \\
&= -5y
\end{aligned}
$$

which is the dot product $\mathbf{a} \cdot \mathbf{b}$ of the vectors dual to the 1-forms a and b.

Example 7.5: The interior product of a 1-form and a 2-form

The interior product of $a = 3x\, dx - y\, dz$ and $c = 4\, dz\, dx + 5\, dx\, dy$ is

$$
\begin{aligned}
a \lrcorner c &= \left(3x\, dx - y\, dz\right) \lrcorner \left(4\, dz\, dx + 5\, dx\, dy\right) \\
&= 12x\, dx \lrcorner dz\, dx + 15x\, dx \lrcorner dx\, dy - 4y\, dz \lrcorner dz\, dx - 5y\, dz \lrcorner dx\, dy \\
&= -12x\, dz + 15x\, dy - 4y\, dx
\end{aligned}
$$

which is the 1-form dual to $-\mathbf{a} \times \mathbf{c}$, where $\mathbf{a}$ and $\mathbf{c}$ are vectors dual to a and c.

The interior product can be related to the exterior product using the star operator. The interior product of arbitrary forms a and b is

$$a \lrcorner b = \star \left(\star b \wedge a \right) \tag{7.48}$$

which can be used to compute the interior product in curvilinear coordinate systems. (This formula shows the metric dependence of the interior product as we have defined it; the interior product is usually defined to be the contraction of a vector with a form, which is independent of any metric.) The interior and exterior products satisfy the identity

$$\alpha = n \wedge \left(n \lrcorner \alpha \right) + n \lrcorner \left(n \wedge \alpha \right) \tag{7.49}$$

where n is a 1-form and α is arbitrary.

The Lorentz force law can be expressed using the interior product. The force 1-form F is

$$F = q \left(E - \mathbf{v} \lrcorner B \right) \tag{7.50}$$

where $\mathbf{v}$ is the velocity of a charge q, and the interior product can be computed by finding the 1-form dual to $\mathbf{v}$ and using the rules given above. F is dual to the usual force vector $\mathbf{F}$. The force 1-form has units of energy, and does not have as clear a physical interpretation as the usual force vector. In this case we prefer to work with the vector dual to F, rather than the 1-form itself. Force, like displacement and velocity, is naturally a vector quantity.

7.7.2 BOUNDARY CONDITIONS

A boundary can be specified as the set of points satisfying $f(x, y, z) = 0$ for some suitable function f. The surface normal 1-form is defined to be the normalized exterior derivative of f,

$$n = \frac{df}{\sqrt{\left(df \lrcorner df \right)}} \tag{7.51}$$

The surfaces of n are parallel to the boundary. Using a subscript 1 to denote the region where $f > 0$, and a subscript 2 for $f < 0$, the four electromagnetic boundary conditions can be written as [20]

$$n \lrcorner \left(n \wedge \left(E_1 - E_2 \right) \right) = 0$$

$$n \lrcorner \left(n \wedge \left(H_1 - H_2 \right) \right) = J_s$$

$$n \lrcorner \left(n \wedge \left(D_1 - D_2 \right) \right) = \rho_s$$

$$n \lrcorner \left(n \wedge \left(B_1 - B_2 \right) \right) = 0$$

where J_s is the surface current density 1-form and ρ_s is the surface charge density 2-form. The form $n \lrcorner (n \wedge \omega)$ is the component of ω which has surfaces perpendicular to the boundary and integrates to the same value as ω over any region lying in the boundary.

7.7.3 SURFACE CURRENT

The action of the operator $n \lrcorner n\wedge$ can be interpreted graphically, leading to a simple picture of the field intensity boundary conditions. Consider the field discontinuity $H_1 - H_2$ shown in Figure 7.17. The exterior product of n and $H_1 - H_2$ is a 2-form with tubes that run parallel to the boundary, as shown in Figure 7.17b. The component of $H_1 - H_2$ with surfaces parallel to the boundary is removed. The interior product $n \lrcorner (n \wedge (H_1 - H_2))$ removes the surfaces parallel to the boundary, leaving only surfaces perpendicular to the boundary, as in Figure 7.17c. Current flows along the lines where the surfaces intersect the boundary. The direction of flow along the lines of the 1-form can be found using the right-hand rule on the direction of $H_1 - H_2$ in region 1 above the boundary.

The field intensity boundary conditions are intuitive: the boundary condition for magnetic field intensity requires that surfaces of the 1-form $H_1 - H_2$ end along lines of the surface current density 1-form J_s, as can be seen in Figure 7.17. The surfaces of $E_1 - E_2$ cannot intersect a boundary at all, so that they must be parallel to the boundary.

Unlike other electromagnetic quantities, J_s is not dual to the vector $\mathbf{J}_s$. The direction of $\mathbf{J}_s$ is parallel to the lines of J_s in the boundary, as shown in Figure 7.17c. (J_s is a twisted differential form, so that under coordinate inversion it transforms with a minus sign relative to a nontwisted 1-form. This property is discussed in detail in Refs. [2, 20, 21]. Operationally, the distinction can be ignored as long as one remains in right-handed coordinates.) J_s is natural both mathematically and geometrically as

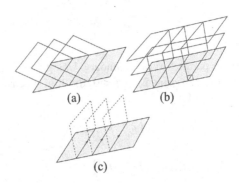

FIGURE 7.17   (a) The 1-form $H_1 - H_2$. (b) The 2-form $n \wedge (H_1 - H_2)$. (c) The 1-form J_s, represented by lines on the boundary. Current flows along the lines.

a representation of surface current density. The expression for current through a path using the vector surface current density is

$$I = \int_P J_s \cdot \left(\hat{\mathbf{n}} \times d\mathbf{l} \right) \tag{7.52}$$

where $\hat{\mathbf{n}}$ is a surface normal. This simplifies to

$$I = \int_P J_s \tag{7.53}$$

using the 1-form J_s. Note that J_s changes sign depending on the labeling of regions one and two; this ambiguity is equivalent to the existence of two choices for n in Equation (7.52).

The following example illustrates the boundary condition on the magnetic field intensity.

Example 7.6: Surface current on a sinusoidal surface

A sinusoidal boundary given by $z - \cos y = 0$ has magnetic field intensity $H_1 = dx$ A above and zero below. The surface normal 1-form is

$$n = \frac{\sin y \, dy + dz}{\sqrt{1 + \sin^2 y}}$$

By the boundary conditions given above,

$$J_s = n \lrcorner (n \wedge dx)$$
$$= \frac{1}{1 + \sin^2 y} (\sin y \, dy + dz) \lrcorner (\sin y \, dy \, dz + dz \, dx)$$
$$= \frac{dx + \sin^2 y \, dx}{1 + \sin^2 y}$$
$$. = dx \text{ A}$$

The usual surface current density vector J_s is $\left(\hat{\mathbf{y}} - \hat{\mathbf{z}} \sin y \right) \left(1 + \sin^2 y \right)^{-1/2}$, which clearly is not dual to dx. The direction of the vector is parallel to the lines of J_s on the boundary,

7.7.4 SURFACE CHARGE

The flux density boundary conditions can also be interpreted graphically. Figure 7.18a shows the 2-form $D_1 - D_2$. The exterior product $n \wedge (D_1 - D_2)$ yields boxes that have sides parallel to the boundary, as shown in Figure 7.18b. The component of $D_1 - D_2$ with tubes parallel to the boundary is removed by the exterior product. The

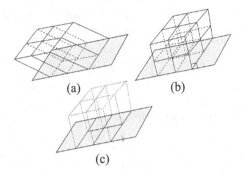

FIGURE 7.18   (a) The 2-form $D_1 - D_2$. (b) The 3-form $n \wedge (D_1 - D_2)$, with sides perpendicular to the boundary. (c) The 2-form ρ_s, represented by boxes on the boundary.

interior product with n removes the surfaces parallel to the boundary, leaving tubes perpendicular to the boundary. These tubes intersect the boundary to form boxes of charge (Figure 7.18c). This is the 2-form $\rho_s = n \rfloor (n \wedge (D_1 - D_2))$.

The flux density boundary conditions have as clear a graphical interpretation as those for field intensity: tubes of the difference $D_1 - D_2$ in electric flux densities on either side of a boundary intersect the boundary to form boxes of surface charge density. Tubes of the discontinuity in magnetic flux density cannot intersect the boundary.

The sign of the charge on the boundary can be obtained from the direction of $D_1 - D_2$ in region 1 above the boundary, which must point away from positive charge and toward negative charge. The integral of ρ_s over a surface,

$$Q = \int_S \rho_s \tag{7.54}$$

yields the total charge on the surface. Note that ρ_s changes sign depending on the labeling of regions one and two. This ambiguity is equivalent to the existence of two choices for the area element dA and orientation of the area A in the integral $\int_A q_s \, dA$, where q_s is the usual scalar surface charge density. Often, the sign of the value of the integral is known beforehand, and the subtlety goes unnoticed.

7.8   CONCLUSION

The primary pedagogical advantages of differential forms are the distinct representations of field intensity and flux density, intuitive graphical representations of each of Maxwell's laws, and a simple picture of electromagnetic boundary conditions. Differential forms provide visual models that can help students remember and apply the principles of electromagnetics. Computational simplifications also result from the use of forms: derivatives are easier to employ in curvilinear coordinates, integration becomes more straightforward, and families of related vector identities are replaced by algebraic rules. These advantages over traditional methods make the calculus of differential forms ideal as a language for teaching electromagnetic field theory.

The reader will note that we have omitted important aspects of forms. In particular, we have not discussed forms as linear operators on vectors, or covectors, focusing instead on the integral point of view. Other aspects of electromagnetics, including vector potentials, Green functions, and wave propagation also benefit from the use of differential forms.

Ideally, the forms-based electromagnetics curriculum would be taught in conjunction with calculus courses employing differential forms. A unified curriculum, although desirable, is not necessary in order for students to profit from the use of differential forms. We have found that because of the simple correspondence between vectors and forms, the transition from vector analysis to differential forms is generally quite easy for students to make. Familiarity with vector analysis also helps students to recognize and appreciate the advantages of the calculus of differential forms over other methods.

We hope that this attempt at making differential forms accessible at the undergraduate level helps to fulfill the vision expressed by Deschamps [4] and others, that students obtain the power, insight, and clarity that differential forms offer to electromagnetic field theory and its applications.

REFERENCES

1. K. F. Warnick and R. H. Selfridge and D. V. Arnold. Teaching Electromagnetic Field Theory Using Differential Forms. *IEEE Trans. Educ.*, 40(1):53–68, 1997.
2. William L. Burke. *Applied Differential Geometry*. Cambridge University Press, Cambridge, 1985.
3. C. Misner and K. Thorne and J. A. Wheeler. *Gravitation*. Freeman, San Francisco, 1973.
4. G. A. Deschamps. Electromagnetics and Differential Forms. *IEEE Proc.*, 69(6):676–696, 1981.
5. C. Nash and S. Sen. *Topology and geometry for physicists*. Academic Press, San Diego, California, 1983.
6. P. Bamberg and S. Sternberg. *A Course in Mathematics for Students of Physics*, volume II. Cambridge University Press, Cambridge, 1988.
7. H. Flanders. *Differential Forms with Applications to the Physical Sciences*. Dover, New York, 1963.
8. Y. Choquet-Bruhat and C. DeWitt-Morette. *Analysis, Manifolds and Physics*. North-Holland, Amsterdam, Rev. edition, 1982.
9. S. Hassani. *Foundations of Mathematical Physics*. Allyn and Bacon, Boston, 1991.
10. Robert Hermann. *Topics in the geometric theory of linear systems*. Math Sci Press, Brookline, MA, 1984.
11. D. Baldomir. Differential forms and electromagnetism in 3-dimensional Euclidean space R^3. *IEE Proc.*, 133(3):139–143, 1986.
12. D. Baldomir and P. Hammond. Global geometry of electromagnetic systems. *IEE Proc.*, 140(2):142–150, 1992.
13. P. Hammond and D. Baldomir. Dual energy methods in electromagnetics using tubes and slices. *IEE Proc.*, 135(3A):167–172, 1988.
14. D. B. Nguyen. Relativistic constitutive relations, differential forms, and the p-compound. *Am. J. Phys.*, 60(12):1137–1147, 1992.

15. N. Schleifer. Differential forms as a basis for vector analysis—with applications to electrodynamics. *Am. J. Phys.*, 51(12):1139–1145, 1983.

16. R. S. Ingarden and A. Jamiolkowksi. *Classical Electrodynamics*. Elsevier, Amsterdam, The Netherlands, 1985.

17. S. Parrott. *Relativistic Electrodynamics and Differential Geometry*. Springer-Verlag, New York, 1987.

18. Walter Thirring. *Classical Field Theory*, volume II. Springer-Verlag, New York, 2nd edition, 1978.

19. J. A. Schouten. *Pfaff's Problem and its Generalizations*. Chelsea, New York, 1969.

20. K. F. Warnick and R. H. Selfridge and D. V. Arnold. Electromagnetic Boundary Conditions Using Differential Forms. *IEE Proc.*, 142(4):326–332, 1995.

21. William L. Burke. Manifestly parity invariant electromagnetic theory and twisted tensors. *J. Math. Phys.*, 24(1):65–69, 1983.

8 Maxwell's Displacement Current

A Teaching Approach Infusing Ideas of Innovation and Creativity

Krishnasamy T. Selvan

Sri Sivasubramaniya Nadar College of Engineering, India

CONTENTS

8.1 INTRODUCTION

In this chapter, which is a thoroughly revised version of [1], we consider the teaching of the displacement current concept, "the essence of one of the most profound discoveries in physics since Newton" [2]. While a typical classroom discussion of this topic reflects "a pedagogical tradition, whose aim is...to help the student understand and feel at home with the current formalism" [3], it is suggested that considering the topic in its rich historical and scientific context of development can provide a more enriching learning experience, for reasons that are discussed throughout this chapter and in particular in Section 8.9. To this end, this chapter attempts to present a more complete story of the displacement current concept.

The chapter mainly focusses on the following themes, the overall objective being a consideration of how they can be employed in the classroom:

- Maxwell's changing approach to science during his 25 years of active research engagement with electromagnetic theory
- The actual context in which Maxwell introduced the displacement current concept
- A discussion of two prominent questions in the literature, namely, whether the displacement current is equivalent to electric current and whether it produces a magnetic field

The chapter is structured as follows: In Section 8.2, we discuss how Maxwell's approach to science was changing as he went on from one significant contribution to the next. Section 8.3 discusses the molecular vortex model, and Section 8.4 presents how Maxwell used that model to introduce the displacement current concept. Section 8.5 considers the typical textbook presentation of the topic. Section 8.6 discusses whether displacement current is equivalent to electric current, while Section 8.7 considers the related question of whether it produces a magnetic field. Sections 8.8 and 8.9 focus on the importance of the term "displacement current" and how a context-based discussion of it can be pedagogically beneficial. Section 8.10 concludes the chapter.

8.2 MAXWELL'S DYNAMIC APPROACH TO SCIENCE

Maxwell's active research engagement with electromagnetic theory spanned a 25-year-period (mid-1850s to 1879). Systematically pursuing his "goal-oriented theoretical endeavor" during this period, Maxwell established a "complete and enduring foundation for electromagnetic theory," among other achievements, by dynamically adjusting his scientific approach to emerging challenges [4]. In this journey, he displayed extraordinary ingenuity in modifying existing methodologies in unforeseen ways [5]. (It is of interest to mention here that Maxwell's contributions were not limited to electromagnetics. For an overview of his other achievements, which were in topics as varied as digital photography and Saturn's rings, the reader may refer to [6].)

In the early stages, Maxwell favored the use of physical *analogies* in preference to purely mathematical or physical theories. He said that while a purely mathematical theory would result in losing "sight of the phenomena to be explained," a physical hypothesis could just be a "partial explanation," and thus be a threat to our discovering facts [7]. At the same time, he hoped that "a mature theory, in which physical facts will be physically explained" would be developed in the future [7].

In his later work [8], perhaps toward developing such a theory, Maxwell "sought to explain physical phenomena mechanically." He held that "when a physical phenomenon can be completely described as a change in the configuration and motion of a material system, the dynamical explanation of that phenomenon is said to be complete." Maxwell had the clue for this approach in William Thomson's work on molecular vortices [9]. Thomson had suggested that "electromagnetism has a mathematical structure analogous to heat conduction," and that "all these phenomena could be understood mechanically" [5]. In his *Treatise* [10], Maxwell said although the "physical relation between electrified bodies" might be conceived as arising either from an action-at-a-distance phenomenon or through the intervening medium, it is only the latter that could facilitate a physical, rather than a purely mathematical, explanation of the relation, and therefore deserved preference. Despite this stand, Maxwell admitted that the two approaches were mathematically equivalent [10]. Going further, he believed "it is a good thing to have two ways of looking at a subject, and to *admit* that there are two ways of looking at it." A real mark of openness!

Although Maxwell tended to take a skeptical stance concerning the mechanical representation toward the end of his career, in the middle period, which was also "the period of intensive innovation" in which "two major innovations in Maxwell's electromagnetic theory – the displacement current and the electromagnetic theory of light – received their initial formulations in the context of the molecular-vortex model" [4], his work displayed a strong commitment to the mechanical approach.

Did Maxwell's commitment to mechanical approach mean he assumed the models to be realistic? Not at all! He carefully distinguished between nature and our abstraction of it [11]:

> ...molecules have laws of their own, some of which we select as most intelligible to us and most amenable to our calculation. We form a theory from these partial data, and we ascribe any deviation of the actual phenomena from this theory to disturbing causes. At the same time, we confess that what we call disturbing causes are simply those parts of the true circumstances which we do not know or have neglected, and we endeavor in future to take account of them. We thus acknowledged that the so-called disturbance is a mere figment of mind, not a fact of nature, and that in natural action there is no disturbance.

To Maxwell, the model thus did not necessarily represent reality. While the usefulness of any model lay in simplifying reality and in possibly leading to a satisfactory theory for explaining and predicting observations, he took care to "stress that no physical explanation could be in perfect correspondence with reality" [12]. Thus, a scientist must remain open to newer experiences and perspectives, or in other words be an "unscrupulous opportunist" as Einstein said [13], so that they can continue to improve their approach to research. After all, "breaking from tradition is necessary for scientific advancement" [14]. As we have briefly traced in this Section, Maxwell is a great example of how openness and adaptability help successful research.

8.3   THE MOLECULAR VORTEX MODEL

Maxwell's primary goal was the construction of a "complete and consistent mechanical model of the electromagnetic field" [3]. During 1861–62, he developed a molecular-vortex representation for this phenomenon, based on Thomson's earlier work on the topic.

In this representation, any medium can be considered to be consisting of innumerable tiny ether cells or vortices that are separated from each other by small particles, acting as idle wheels, as shown in Figure 8.1. Each molecular vortex was modeled for illustrative purposes as hexagonal in cross-section and as a "blob of elastic material" [3]. In a conducting material, under the influence of a magnetic field, each of these vortices rotates in the same sense. The axis of the vortex rotation corresponds to the direction of the magnetic field, while the speed of rotation corresponds to its magnitude. The rotation of the vortices will make the idle wheels move from vortex to vortex, constituting a conduction current. In insulators, including free space, the idle wheels could not leave their place, with the implication that any movement on their part would result in a distortion of the corresponding cells, as they are elastic. These distortions represented an electric field. By using this difference between conductors and insulators, Maxwell was able to account for the accumulation of charges at the boundaries between insulators and conductors [15].

This molecular vortex model was able to account for all electromagnetic phenomena known at the time. For various examples, the reader can refer to the very insightful [3,16]. We will briefly consider the specific case of displacement current in the context of the molecular vortex model in the next section.

8.4   HOW MAXWELL INTRODUCED DISPLACEMENT CURRENT USING MECHANICAL MODEL

Maxwell's first mention of the term "electric displacement" appears to have been made in [8]. He "had long been aware that…Ampere's circuital law in differential form had restricted applicability"– that is, only to closed circuits – "and that was a

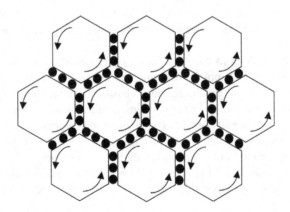

FIGURE 8.1   Maxwell's vortices and idle-wheel particles. (Redrawn from [4])

matter of concern to him." It was toward rectifying this law, and within the overall primary goal of developing a "complete and consistent mechanical model of the electromagnetic field" [3] that, in January 1862, he proposed and used the concept of displacement current [4]. To visualize how Maxwell introduced displacement current in the context of the molecular vortex model, let us refer to Figure 8.2 that shows the section of a thick wire in which a thin gap forms a capacitor [3]. The wire actually forms a loop (except for the gap capacitor) and is connected to a voltage source (not shown in the figure). Surfaces AA and BB, respectively, represent the left and right edges of the gap capacitor. The gap is filled with a dielectric, which could also be free space. When a current flows through the wire, the idle wheels move in the conductor. In so moving, the vortices are made to rotate about their axes as shown by the arrows, with a speed proportional to the magnitude of the magnetic field created. (The magnetic field inside a conductor linearly increases with radius.) The length of the arrows represents this variation.

If the particles move toward the right, they accumulate at edge AA, as they cannot move further into the dielectric. Thus, a net positive charge builds up on the surface AA. From surface BB, the particles will move away, resulting in a net negative charge there. The rows of vortices within the wire on either side of the gap drive the vortices within the dielectric material through the stationary particles (as within the dielectric they are not free to move) at the surfaces AA and BB. This results in the distortion of the vortices within the dielectric. Maxwell suggested that these distortions represent electric field. When the electric field is varied, for example by an alternating source connected to the wire, "there are small changes in the positions of the electric particles in the insulating medium, or vacuum, and so there are small electric currents associated with ... the *displacement* of the electric particles from their equilibrium positions" [16]. Thus, to Maxwell, the electric displacement vector **D** referred to "the measure of electric polarization, whether in a dielectric" or in vacuum [17].

In Maxwell's words, then, "the true electric current J_T, that on which the electromagnetic phenomena depend, is not the same thing as J_C, the current of conduction, but that the time-variation of **D**, the electric displacement, must be taken into account

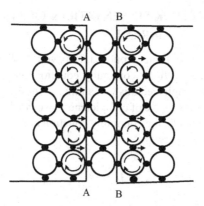

FIGURE 8.2 Molecular vortex model for the capacitor formed by a thin gap in a thick wire. (Redrawn based on [3])

in estimating the total movement of electricity," so that we can write $\mathbf{J_T} = \mathbf{J_C} + \mathbf{J_D}$ [2]. Here $\mathbf{J_D}$ is the displacement current density. Thus, it appears that Maxwell suggested that the "electric displacement...is a movement of electricity in the *same sense* as" the normal "movement of electricity" [10].

Maxwell justified the introduction of the concept of current in a dielectric by saying [10]:

> We have very little experimental evidence to the direct electromagnetic action of currents due to the variation of electric displacement in dielectrics, but the extreme difficulty of reconciling the laws of electromagnetism with the existence of electric currents which are not closed is one reason among many why we must *admit* the existence of transient currents due to the variation of displacement.

Thus, Maxwell was aware of the limitations within which he thought the new term had to be introduced. He was quite clear about what the model meant and what it did not. He wrote [16]: "The conception of a particle having its motion connected with that of a vortex by perfect rolling contact may appear somewhat awkward. I do not bring it forward as a mode of connection existing in nature....It is however a mode of connection which is mechanically conceivable and it serves to bring out the actual mechanical connections between known electromagnetic phenomena." Although Maxwell later reformulated the theory without any special assumptions about the medium [8], his most significant innovations, including the introduction of displacement current concept, were made within the context of the mechanical model, as noted earlier.

It is of interest to note that even some of Maxwell's closest colleagues were extremely skeptical of the displacement current concept. Thomson, for example, "simply didn't believe that such a thing as displacement current could exist," especially in "the nothingness of a vacuum" [18]. But then to Maxwell, as we remarked above, the goal was a comprehensive working, not necessarily realistic, mechanical model that could explain the continuity of current flow in open circuits.

8.5 TYPICAL TEXTBOOK PRESENTATION OF MAXWELL'S DISPLACEMENT CURRENT

With a comprehensive model thus proposed, Maxwell modified Ampere's law so as to ensure its consistency with the continuity equation and to generalize it to open circuits. The way he accomplished this was similar to the account in most text books; for example [19–21]. For the sake of completeness, we quickly review the steps here.

For electromagnetic fields, the continuity equation that represents the principle of charge conservation is given by

$$\nabla \cdot \mathbf{J_C} = -\frac{\partial \rho}{\partial t} \tag{8.1}$$

where ρ is the charge density.

Ampere's original law is

$$\nabla \times \mathbf{H} = \mathbf{J}_C \tag{8.2}$$

It is worth noting here that though Equation 8.2 is usually called Ampere's law, it was Maxwell who originally wrote it down [17].

Taking the divergence on either side of Equation 8.2,

$$\nabla \cdot \nabla \times \mathbf{H} = \nabla \cdot \mathbf{J}_C = 0 \tag{8.3}$$

Ampere's original law thus being inconsistent with the continuity equation, it does not hold for time-varying fields. Maxwell introduced the displacement–current term $\mathbf{J_D} = \dfrac{\partial \mathbf{D}}{\partial t}$ to modify Ampere's law:

$$\nabla \times \mathbf{H} = \mathbf{J}_C + \frac{\partial \mathbf{D}}{\partial t} \tag{8.4}$$

It can be shown that Equation 8.4 obeys the continuity equation and also remains valid for open circuits such as a circuit containing an ac source and a capacitor.

When we consider Equation 8.4, or for that matter any of the other Maxwell's equations, we generally understand that the right-hand side represents the *cause* or the *source*, while the left-hand side represents the *effect*. Thus, $\mathbf{J_D}$ should represent a source term. But when we see that $\mathbf{J_D}$ is changing electric field, and thus should actually be an effect, we wonder if Maxwell indeed considered $\mathbf{J_D}$ to be a source term and if so, was he right in doing that. Subsequent discussions will throw some light on this and related questions.

8.6 WAS MAXWELL JUSTIFIED IN TREATING DISPLACEMENT CURRENT AS EQUIVALENT TO ELECTRIC CURRENT?

It is helpful to recollect the Coulomb gauge, a condition frequently employed in electromagnetics, as we begin to discuss this question. If $\mathbf{A}$ is magnetic vector potential, this condition is given by

$$\nabla \cdot \mathbf{A} = 0 \tag{8.5}$$

Given that electromagnetic waves travel with a finite velocity, we would expect the potential to be retarded too. However, the Coulomb gauge predicts instantaneous potentials, though it predicts the final fields correctly.

It has been suggested [22] that Maxwell's treating the displacement current as being equivalent to current is attributable to his "constant choice of the Coulomb gauge for the potentials." In order to understand this better, let's start with the inhomogeneous wave equations for scalar potential Φ and vector potential $\mathbf{A}$, explicitly showing the displacement current term:

$$-\nabla^2\Phi - \frac{\partial}{\partial t}\left(\nabla\cdot\mathbf{A}\right) = \frac{\rho}{\varepsilon_0} \tag{8.6}$$

$$-\nabla^2\mathbf{A} + \nabla\left(\nabla\cdot\mathbf{A}\right) = \mu_0\left[\mathbf{J}_C + \frac{\partial\mathbf{D}}{\partial t}\right]. \tag{8.7}$$

Employing the Coulomb gauge, Equations 8.6 and 8.7 become

$$-\nabla^2\Phi = \frac{\rho}{\varepsilon_0} \tag{8.8}$$

$$-\nabla^2\mathbf{A} = \mu_0\left[\mathbf{J}_C + \frac{\partial\mathbf{D}}{\partial t}\right] \tag{8.9}$$

While Equation 8.8 is Poisson's equation, Equation 8.9 "has the appearance of a Poisson's equation for the vector potential, with a source term that is the sum of the conduction current density and the displacement current" [22]. Maxwell thus "preferred to have scalar and vector potentials satisfy Poisson-like equations with source terms, the charge density as source for the equation for the scalar potential and the total current for the vector potential. It never bothered him that his total current was not really a source term, but contained an initially unknown displacement current" [1,23]. According to Jackson, "it is in this sense, and within the framework of the Coulomb gauge, that Maxwell could insist on the reality of the displacement current as a contribution to the total current" [23].

Roche [24], while appreciating that "in the Coulomb gauge," which is frequently used in advanced electromagnetism, "the displacement current can, in general, be considered to be equivalent to an electric current," added that the "Coulomb gauge is not a physical gauge and the equivalence of $\frac{\partial\mathbf{D}}{\partial t}$ to a current is always purely fictional" [25]. Responding to this observation, Jackson observed that "the choice of gauge is purely a matter of convenience. The Coulomb gauge is no more or less physical than any other. It is convenient for some problems, inconvenient for others. The fields are the reality" [26]. It has also been noted also by Yaghjian that there is nothing mathematically wrong with the Coulomb gauge [1].

So, is there a simple, straight answer to the question of whether the displacement current can be considered as equivalent to electric current and thus as a source term? No, as the answer would depend on whether one considers the use of the Coulomb gauge as an acceptable technique. Views can differ here and so can the answer.

The natural next question is on whether $\mathbf{J}_D$ produces magnetic fields.

8.7 DOES THE DISPLACEMENT CURRENT PRODUCE A MAGNETIC FIELD?

Arguments have been proposed both in support of, and in opposition to, the statement that the displacement current produces a magnetic field. We will review these arguments briefly in this section.

8.7.1 No, It Doesn't!

It has been shown by several authors that for slowly changing fields, one cannot observe the magnetic field caused by the displacement current.

Purcell [27] showed this by considering the case of a slowly discharging capacitor. Since the electric field is slowly diminishing, it can therefore be practically considered to be an electrostatic field. Its curl is thus practically zero. By implication, the curl of the displacement–current density must also be nearly zero.

Starting with

$$\mathbf{J_D} = \frac{\partial \mathbf{D}}{\partial t} = \varepsilon \frac{\partial \mathbf{E}}{\partial t} \tag{8.10}$$

and taking the curl of either side,

$$\nabla \times \mathbf{J_D} = \nabla \times \varepsilon \frac{\partial \mathbf{E}}{\partial t} = \varepsilon \frac{\partial}{\partial t}\left(\nabla \times \mathbf{E}\right) = -\varepsilon \frac{\partial^2 \mathbf{B}}{\partial t^2} \tag{8.11}$$

where we have made use of Faraday's law. The curl of $\mathbf{J_D}$, as predicted by Equation 8.11, will be negligible for sufficiently slowly varying, or quasistatic, fields. The implication of the displacement–current density having insignificant curl is that it's divergence, following Helmholtz's theorem, should be nonzero. Thus, $\mathbf{J_D}$ should comprise radially directed currents, whose magnetic field should be zero by symmetry. It follows that for slowly varying fields the displacement current does not produce significant magnetic fields.

Bartlett [28] and French and Tessman [29] showed that the magnetic field for slowly varying fields can be estimated without including displacement current when the Biot-Savart law is employed.

Rosser [30] showed that the displacement current does not produce magnetic field in empty space by starting with the inhomogeneous wave equation for the vector potential and by using the Coulomb gauge:

$$\nabla^2 \mathbf{A} - \mu_0 \varepsilon_0 \frac{\partial^2 \mathbf{A}}{\partial t^2} = -\mu_0 \mathbf{J_C} + \mu_0 \varepsilon_0 \nabla \frac{\partial \phi}{\partial t} \tag{8.12}$$

The right-hand side of Equation 8.12 can be written as

$$-\mu_0 \mathbf{J_C} + \mu_0 \varepsilon_0 \nabla \frac{\partial \phi}{\partial t} = -\mu_0 \left(\mathbf{J_C} - \varepsilon_0 \nabla \frac{\partial \phi}{\partial t} \right) = -\mu_0 \mathbf{J_t} \tag{8.13}$$

where $\mathbf{J_t}$, called the transverse current density, can be calculated from the conduction-current density, $\mathbf{J_C}$, alone and the scalar potential. Since the displacement current in empty space does not appear in this equation, "it seems pointless to make statements such as *a changing electric field produces a magnetic field*" [30]. It should be noted that in this approach, the displacement–current term is mathematically eliminated from the wave equation by using the scalar potential.

8.7.2 YES, IT DOES!

It has been shown that the application of the Maxwell–Ampere law for the estimation of magnetic field requires only the real currents for certain contours, while the inclusion of displacement current is warranted to obtain the correct result for arbitrary choices of contours [22,28,31]. Jackson [22] demonstrated this approach by considering Maxwell–Ampere law around two loops: loop A and loop B, as in Figure 8.3. While the application on loop A leads to a result that relates the magnetic field only to the conduction current, the application on loop B requires the inclusion of only the displacement current so that the result can be consistent with that of loop A.

In a further discussion using an alternate method, Jackson showed that the displacement current is the only source of magnetic field within a charging capacitor. With reference to Figure 8.3, a current, $I(t)$, flows along the negative z direction. It brings a total charge of $Q(t)$ to the plate at $z = 0$, and removes an equal amount of charge from the plate at $z = d$. Assuming the static limit, and neglecting fringing (as $a \gg d$), $Q(t)$ is uniformly distributed over the inner side of the positive plate. The surface charge density there is thus $\sigma(t) = \dfrac{Q(t)}{\pi a^2}$. The electric displacement D$_z$ is uniform over the entire volume between the plates and is directed along negative z. Thus:

$$D_z = -\sigma(t) = -\frac{Q(t)}{\pi a^2} \tag{8.14}$$

The displacement current is therefore given by

$$\mathbf{J_D} = \frac{\partial \mathbf{D}}{\partial t} = -\hat{z}\frac{I(t)}{\pi a^2} \tag{8.15}$$

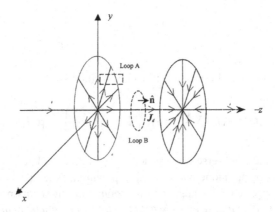

FIGURE 8.3 A charging capacitor. Each circular plate is of radius a. The plates are separated by a distance d, such that $d \ll a$.

Since there is no conduction current between the plates, the Ampere–Maxwell law for the situation is

$$\nabla \times \mathbf{H} = -\hat{z}\frac{I(t)}{\pi a^2} \tag{8.16}$$

From Equation 8.16, and recognizing that the displacement current causes an azimuthal magnetic field, the magnetic field for $0 < \rho < a$ and $0 < z < d$ can be shown to be [22]

$$H_\phi = -\frac{I(t)\rho}{2\pi a^2} \tag{8.17}$$

The magnetic field in the region between the capacitor plates thus depends only on the existence of Maxwell's displacement current.

Zapolsky [32] argued that Rosser's argument presented in Section 8.7.1 - that no part of the displacement current gives rise to a magnetic field (as the term does not appear in the wave equation for vector potential) - is essentially a semantic argument. This can be explicitly shown as follows [1,23].

Let us again start with the inhomogeneous wave equation for the vector potential, Equation 8.12:

$$\nabla^2 \mathbf{A} - \mu_0 \varepsilon_0 \frac{\partial^2 \mathbf{A}}{\partial t^2} = -\mu_0 \mathbf{J_C} + \mu_0 \varepsilon_0 \nabla \frac{\partial \phi}{\partial t} \tag{8.12}$$

The scalar potential can be written as

$$\nabla \phi = -\frac{\partial \mathbf{A}}{\partial t} - \mathbf{E} \tag{8.18}$$

Differentiating either side of Equation 8.18 and multiplying by $\mu_0 \varepsilon_0$,

$$\mu_0 \varepsilon_0 \nabla \frac{\partial \phi}{\partial t} = -\mu_0 \varepsilon_0 \frac{\partial^2 \mathbf{A}}{\partial t^2} - \mu_0 \frac{\partial \mathbf{D}}{\partial t} \tag{8.19}$$

substituting the above equation into Equation 8.12,

$$\nabla^2 \mathbf{A} - \mu_0 \varepsilon_0 \frac{\partial^2 \mathbf{A}}{\partial t^2} = -\mu_0 \mathbf{J_C} - \mu_0 \varepsilon_0 \frac{\partial^2 \mathbf{A}}{\partial t^2} - \mu_0 \frac{\partial \mathbf{D}}{\partial t} \tag{8.20}$$

Thus,

$$\nabla^2 \mathbf{A} = -\mu_0 \left(\mathbf{J_C} + \frac{\partial \mathbf{D}}{\partial t} \right) \tag{8.21}$$

It can thus be seen that Rosser's Equation 8.13 in terms of transverse current density has actually hidden away the displacement current! His equation is in fact equivalent to Equation 8.20, which involves $\dfrac{\partial \boldsymbol{D}}{\partial t}$.

According to Roche [24], the investigations by Purcell, Zapolsky, and Bartlett demonstrated that "the displacement current of a rapidly changing induced electric field will generate a significant magnetic field." When one has rapidly changing fields, the induction effect of changing electric field producing magnetic field is thus observable.

8.7.3   REFLECTING ON THE QUESTION

Presenting the class with both sides of the argument discussed above and asking the students to come up with their points of view on whether the displacement current produces a magnetic field will enable them to reflect on the topic and think critically. It can encourage class discussions that can in turn hone the students' higher thinking skills [33].

Do the discussions in Sections 8.7.1 and 8.7.2 contradict each other? Not really, if we consider that the displacement current, rather than being a "real" current, is only a "way of describing how the change in electric field passing through a particular area can give rise to a magnetic field, just as a current does" [18].

8.8   IMPORTANCE OF RETAINING THE TERM *DISPLACEMENT CURRENT*

Considering the preceding discussions, is it appropriate to call $\mathbf{J_D}$ displacement *current*? We will first present views on this question from the literature and then consider the question from a pedagogical perspective.

Warburton [34], while agreeing that the displacement current was a reasonable concept when it was introduced, felt that the name had lost its relevance as the concept of ether filling all space including vacuum is no longer in use. He also suggested that this term might mislead students. Rosser [30] suggested that "a lot of confusion about the role of the displacement current in empty space might be avoided, if it were called something else that did not include the term current. If a name is needed, it could be called the Maxwell term in honor of the man who first introduced it."

Closely agreeing with the above views, Roche [24] stated that while the term $\dfrac{\partial \boldsymbol{D}}{\partial t}$ is "enormously important," it can in no sense be "described as an electric current" in the case of a uniformly charging capacitor. He also proceeded to suggest that this term should not even be mentioned to undergraduates.

Thus, while it has on the one hand been acknowledged, as by French and Tessman [29], that without the introduction of the displacement current, "the treatment of electromagnetic waves would be absurdly complicated if the fields were always referred back to the motions of real charges," there on the other has also been the suggestion that the name be changed to something else.

But the name carries with it a conceptually and historically rich story with it. If we dropped the name, we would be eventually dropping the story as well! While even now we may not be discussing the evolution of the term in great detail, at least there exists some scope for us to link it with the context in which it was introduced. If the name were to be changed, we would very likely miss out on a great opportunity to discuss the way science progresses and creative geniuses go about their work. Thus, there is a lot of historical and pedagogical significance in this term and therefore it is not that we should just retain the term but seriously consider making appropriate use of it in class to inspire reflective thinking.

8.9 A TEACHING APPROACH INFUSING SCIENTIFIC SPIRIT

While a text-book based discussion of the displacement current does provide the "motivation and rationalization of the current state of affairs in electromagnetic theory" [3] and thus does capture the essential aspects of the *final* Maxwell–Ampere law, this modern formulation differs in "very significant ways" from Maxwell's original formulation, as has been briefly discussed in this chapter. The *development* of the displacement current concept, as outlined in this chapter, actually represents a rich story, accommodating the highest level of scientific and philosophical content. Discussing at least some aspects of this story, as can be constructed from Maxwell's own works and from scholarly contributions subsequent to the introduction of the displacement current concept, can have very desirable educational consequences and should be seriously considered.

There are important reasons for suggesting such an approach. The purpose of higher education, as discussed in Chapters 1 and 2, is not just to transmit "ways of doing" but to develop "reflective practitioners" who can "engage critically with their discipline and make informed judgements" [35]. This requires that the students are helped to develop "an appreciation of the significance of the scientific approach" [36]. Considering that development of science does not often occur in a simplistic way, as to no scientist ideas occur "suddenly and out of nowhere" [18] this would in turn warrant that scientific theories be presented along with the context in which they were developed, the methodology adapted, and an appropriate historical account. Such an approach can help students "to experience the *how* of scientific enquiry, rather than merely being exposed to what is known about and by science" [37]. In this way, an educator can attempt to inculcate scientific spirit in the students by helping them challenge their "image of the certainty of scientific knowledge" and by encouraging them to try to find "place for their own interpretation" [37]. A consequence may be that the students would develop an understanding of the process of knowledge creation. Given the close relationship between knowledge creation and innovation [38], the approach may in turn help with inculcating the qualities of innovation and creativity in them.

Given the well-known pressures of higher education in these times, as discussed in Chapter 2, would it be worth running a session on the topic? The answer seems to be yes, as "the textbook-centered presentation of the finished products of science" cannot provide students with an understanding of "how scientific knowledge comes into being and what is required for the body of scientific knowledge to undergo change" [39]. Thus, it appears that an hour or so of time spent for discussing these ideas is time well-spent and can contribute in our efforts toward developing more scientifically tempered, independent thinking students.

While there can be different perspectives on how the topic of displacement current can be discussed in the class, a suggested approach is as follows:

1. Discuss the context of the development of displacement current concept by Maxwell, and present the textbook account (Sections 8.2 through 8.5);
2. Discuss how the introduction of the term led to most consequential predictions;
3. Discuss the two fundamental questions on the nature of displacement current (Sections 8.6 and 8.7)
4. Present both the positive and negative criticism of the term (Section 8.8)
5. By not taking sides in the discussions above, point out that different perceptions in science pursuit are possible, and in fact desirable, and invite students' reflections.

A power-point presentation more or less following the above approach is available in [40]. An alternative to a lecture session by the educator might be handing out a reading assignment to students, with the follow-up requirement of either an oral presentation or written submission on the key take-aways from the story of the development of Maxwell's displacement current.

The presentation as in [40], or a variation of it, has been used in the author's classes and at several technical meetings, with surveys conducted at some of them. Positive feedback has been received [41,42]. While it is unlikely that an entire class or audience will find the approach appealing – of course this is true of any approach – there will always be students who will find it interesting and useful. Teacher enthusiasm, as discussed in Chapter 2, can of course play a key role here. It has been suggested that apart from sprinkling the ideas of innovation, creativity and openness, such a presentation may entail certain additional benefits, such as acting as a motivator and making the lecture interesting [43,44].

8.10 CONCLUSION

The displacement–current concept, described by Einstein as "closest to...a genuine, useful, profound theory...built purely speculatively" [45], is at the very heart of Maxwell's theory, in the sense that it led to predictions of fundamental importance. This term carries with it a wealth of scientific and philosophical significance. As such, this theory appears to be a fertile pedagogical ground for inspiring students to contemplate about innovation and creativity. When an effort is made to present to the students the term along with its context and development, it will have the potential to have positive effects in the long run, consistent with our ideas of education that lay increasing importance on creativity and innovation and higher-level skills [46,47].

As this approach gives a good idea to students concerning the process of scientific enquiry, it can perhaps also facilitate "successful learning of science" [37].

Additional educational resources related to this chapter can be found at the IEEE AP-S Resource Center, https://resourcecenter.ieeeaps.org/.

ACKNOWLEDGMENTS

The author is grateful to Arthur D. Yaghjian for his critical comments on the manuscript. Suggestions from Karl F. Warnick helped refine the presentation in the chapter.

REFERENCES

1. K. T. Selvan, "A revisiting of scientific and philosophical perspectives on Maxwell's displacement current," *IEEE Antennas and Propagation Magazine*, vol. 51, no. 3, pp. 36–46, June 2009.
2. A. D. Yaghjian, "Reflections on Maxwell's treatise (Invited Paper)," *Progress In Electromagnetics Research*, Vol. 149, 217–249, 2014.
3. D. Siegel, "The origin of the displacement current," *Historical Studies in the Physical and Biological Sciences*, vol. 17, no. 1, 1986, pp. 99–146.
4. D. M. Siegel, *Innovation in Maxwell's electromagnetic theory*, New York, Cambridge University Press, 1991.
5. G. Hon an B.R. Goldstein, "*Maxwell's methodological odyssey in electromagnetism*," *8th Quadrennial Fellows Conference*, Pittsburgh Centre for Philosophy of Science, Lund University, Sweden, July 11, 2016.
6. T. K. Sarkar, R. J. Mailloux, A. A. Oliner, M. Salazar-Palma and D. L. Sengupta, *History of wireless*, New York, Wiley, 2006.
7. J. C. Maxwell, "On Faraday's lines of force," *Transactions of the Cambridge Philosophical Society*, vol. 10, Part I, 1864, pp. 27–65 (read December 10, 1855, and February 11, 1856).
8. J. C. Maxwell, "A dynamical theory of electromagnetic field," *Proceedings of the Royal Society of London*, 1864, pp. 531–536.
9. A. F. Chalmers, "Maxwell, mechanism, and the nature of electricity," *Physics in Perspective*, vol. 3, 2001, pp. 425–438.
10. J. C. Maxwell, *A treatise on electricity and magnetism*, Volume 1, 3rd Edition, Mineola, NY, Dover, 1954 (originally published by Clarendon Press in 1891).
11. P. M. Heimann. "Maxwell, Hertz, and the nature of electricity." *Isis*, vol. 62, no. 2, 1971, pp. 149–157. *JSTOR*, www.jstor.org/stable/229238. Accessed Nov. 26 2020.
12. P. M. Heimann, "Maxwell and the modes of consistent representation," *Archives for History of Exact Sciences*, vol 6, no. 3, January 1970, pp. 171–213.
13. P. Feyerabend, *Against Method*, London: Verso, 1988.
14. S. Koppman and E. Leahey, "Who moves to the methodological edge? Factors that encourage scientists to use unconventional methods," *Research Policy*, Vol. 48, 9, 2019.
15. J. C. Maxwell, *A treatise on electricity and magnetism*, Volume 2, 3rd Edition, Mineola, NY, Dover, 1954 (originally published by Clarendon Press in 1891).
16. M. Longair, *Theoretical concepts in physics: An alternative view of theoretical reasoning in physics*, Cambridge University Press, 3rd ed., Cambridge, 2020
17. A. D. Yaghjian, "Maxwell's definition of electric polarization as displacement," *Progress In Electromagnetics Research M*, Vol. 88, 65–71, 2020.

18. J.C. Rautio, *"The long road to Maxwell's equations,"* IEEE Spectrum, December 2014

19. D. K. Cheng, *Field and wave electromagnetics*, 2nd ed., Addison-Wesley, Boston, 1989

20. F. T. Ulaby, *Fundamentals of applied electromagnetics*, 5th rd., Upper Saddle River NJ, Prentice-Hall, 2007.

21. W. H. Hayt, Jr. and J.A. Buck, *Engineering electromagnetics*, 7th ed., McGraw-Hill, Singapore, 2006.

22. J. D. Jackson, "Maxwell's displacement current revisited," *European Journal of Physics*, **20**, 1999, pp. 495–499.

23. J. D. Jackson, *Classical Electrodynamics*, 3rd Edition, New York, John Wiley & Sons, 1999.

24. J. Roche, "The present status of Maxwell's displacement current," *European Journal of Physics*, vol. 19, 1998, pp. 155–166.

25. J. Roche, "Reply to J.D. Jackson's Maxwell's displacement current revisited," *European Journal of Physics*, vol. 21, no. 4, 2000, pp. L27–L28.

26. J. D. Jackson, "Reply to Comment by J Roche on 'Maxwell's displacement current revisited'," *European Journal of Physics*, **21**, 2000, pp. L29–L30.

27. E. M. Purcell, *Electricity and magnetism*, New York, McGraw-Hill, 1985, pp. 324–330.

28. D. F. Bartlett, "Conduction current and the magnetic field in a circular capacitor," *American Journal of Physics*, vol. 58, no. 12, December 1990, pp. 1168–1172.

29. A. P. French and J. R. Tessman, "Displacement currents and magnetic fields," *American Journal of Physics*, vol. 31, no. 3, March 1963, pp. 201–204.

30. W. G. V. Rosser, "Does the displacement current in empty space produce a magnetic field?," *American Journal of Physics*, vol. 44, no. 12, December 1976, p. 1221–1223.

31. A. J. Dahm, "Calculation of the displacement current using the integral form of Ampere's law," *American Journal of Physics*, vol. 46, 12, December 1978, p. 1227.

32. H. S. Zapolsky, "Does charge conservation imply the displacement current?" *American Journal of Physics*, vol. 55, no. 12, December 1987, p. 1140.

33. https://citl.illinois.edu/citl-101/teaching-learning/resources/teaching-strategies/questioning-strategies, accessed November 21, 2020

34. F. W. Warburton, "Displacement current, a useless concept," *American Journal of Physics*, vol. 22, 1954, pp. 299–305.

35. C. Star and S. Hammer, "Teaching generic skills: eroding the higher purpose of universities, or an opportunity for renewal?," *Oxford Review of Education*, vol. 34, no. 2, 2008, pp. 237–251

36. M. A. B. Whitaker, "History and quasi-history in physics education – Part 1," *Physics Education*, vol. 14, 1979, pp. 108–112.

37. A. R. Irwin, "Historical case studies: Teaching the nature of science in context," *Science Education*, vol. 84, no. 1, pp. 5–26, 2000.

38. S. Popadiuk and C. W. Choo, "Innovation and knowledge creation: How are these concepts related?," *International Journal of Information Management*, vol. 26, no. 4, 2006, pp. 302–312.

39. D.W. Vanderlinden, "Teaching the content and context of science: the effect of using historical narratives to teach the nature of science and science content in an undergraduate introductory geologyss course," PhD dissertation, Iowa State University, 2007.

40. K.T. Selvan, "A story of Maxwell's displacement current," *Forum for Electromagnetic Research Methods and Application Technologies (FERMAT)*, vol. 12, 2015. Available at: https://www.e-fermat.org/files/communication/Selvan-MUL-APCAP2015-Vol12-Nov_Dec-020%20A%20Story%20of%20Maxwells%20Displacement.....pdf

41. K. T. Selvan and S. R. Rengarajan, *"Teaching-in-context of Maxwell's displacement current: What do professors and students perceive?,"* 2010 *IEEE Antennas and Propagation Society International Symposium*, Toronto, ON, 2010, pp. 1–4.

42. K. T. Selvan and L. Ellison, "Incorporation of historical context into teaching: Student perception at the University of Nottingham Malaysia Campus," *IEEE Antennas and Propagation Magazine*, vol. 49, no. 5, October 2007, pp. 161–162.

43. K. T. Selvan and P. F. Wahid, *"Teaching electromagnetic theory: Beyond a focus on applications,"* Proc. *2015 IEEE 4th Asia-Pacific Conference on Antennas and Propagation (APCAP)*, Kuta, 2015, pp. 245–246

44. K. T. Selvan, *"The development of Maxwell's displacement current concept,"* 2009 *Applied Electromagnetics Conference (AEMC)*, Kolkata, 2009, pp. 1–2.

45. D. P. Gribanov, *Albert Einstein's philosophical views and the theory of relativity*, Moscow, Progress, 1987.

46. L. Jamieson, *"Engineering education in a changing world,"* IEC *DesignCon Conference*, January 2007.

47. V. K. Arora and L. Faraone, "21st century engineer-entrepreneur," *IEEE Antennas and Propagation Magazine*, vol. 45, no. 5, October 2003, pp. 106–114.

9 Teaching Electromagnetic Waves to Electrical Engineering Students
An Abridged Approach

Javier Bará Temes

Technical University of Catalonia, Spain

CONTENTS

9.1 INTRODUCTION: THE CURRICULUM BACKGROUND

The rapidly increasing amounts of knowledge and of new technology development are continually pushing back basic, strongly mathematical subjects like electromagnetic theory, solid-state physics, and mathematics itself in engineering curricula. This pressure comes both from curricula designers and from students; while the former have to make room to accommodate new or augmented subjects and new abilities beyond those of "applying knowledge of mathematics, science, and engineering" [1], the latter are reluctant to cope with the long and difficult period of mathematical and formal study required before arriving at practical problem-solving. On the other hand, the mathematical beauty and perfect self-consistency of electromagnetic theory after Maxwell's equations have converted its teaching to a paradigmatic exercise of mathematical derivation of one property after another. While this approach may be intellectually satisfactory when a thorough understanding of the subject is pursued, it is not the best way to direct the efforts of a student who is introduced for the first time to electromagnetic wave phenomena and who is, furthermore, urged to "identify, formulate, and solve engineering problems" [1]. Therefore, the author's point of view is that of emphasis on the student learning process rather than on loyalty to mathematical formality.

The situation is worse in short curricula. In the Spanish university system there are, for a given professional engineering activity (electronic, telecommunication, mechanical, civil engineering, etc.), two different "engineer" degrees: the short curriculum one ("technical engineer," three years) and the long curriculum one ("senior engineer," five years). There is the possibility of completing the long one after the short one with an additional 2 to 3 year study period, this length depending on the particular university.

The "Technical Engineer in Telecommunications" degree is formed by 90% scientific and technical subjects and is therefore, from the point of view of a student's effort devoted to these contents, comparable to a U.S. college bachelor's program. The three-year curriculum, often structured in six semesters, has an introductory physics course (90 lecturing hours) in the first year. The course level is similar to that of Tipler's textbook [2], and half of its contents is devoted to electricity and magnetism topics, including Faraday's induction law. That is, students get a working knowledge of Maxwell's equations in integral form, except for the current displacement term. At the same time, during the first and second semesters the student takes standard courses on calculus, linear systems, electronics, and computers.

The course on electromagnetic waves is taken by the student in his/her third semester and has 56 classroom hours distributed in 14 weeks. Of these hours, 35 correspond to expositive lectures (2.5 a week), while the remaining 21 correspond to professor tutored, collaborative, problem solving (1.5 a week). The course also requires from the student an additional 80-hour personal study, and it is followed by (1) a course on guided wave communication systems, which begins with transmission lines and includes cable and fiber optics communication systems; (2) a course on antennas, propagation, radio relay links, and cellular telephony basics; and (3) a course on radio and microwave communication systems, including both passive and active components. The scope of the courses, which are oriented to understanding actual systems, leaves little time for the analysis of electromagnetic phenomena, and puts pressure on the electromagnetic waves course to reach as far as possible in applications. Under these circumstances, a standard electromagnetic theory course based on Maxwell equations would require from the student a substantial mathematical effort at the expense of relevant and useful technical results.

9.2 COURSE OUTLINE

The course objective is to introduce the students to the full vector description of spherical and plane, time-harmonic, electromagnetic waves; their generation from an oscillating dipole; their polarization and interference properties; their reflection and refraction at conducting and dielectric surfaces; their diffraction by an aperture; and their relationship to basic geometrical and physical optics. The course is based on the postulation of the fields produced by an accelerated charge in the non-relativistic, small velocity limit, which can be viewed as a generalized form of Coulomb's law. It should be noted again that transmission lines are the starting point of a subsequent course on guided wave communication systems that are introduced from a distributed L-C circuit model [3].

In the following paragraphs, the main points of the course syllabus are sketched and omitted in the case of material which can be introduced in a form independent of the course approach (losses in a dielectric, wave attenuation, radiation from elementary wire antennas, etc.). More details can be found in [4].

9.2.1 PART I: A GENERALIZED COULOMB-AMPÈRE LAW

After a short review of spherical and cylindrical coordinates, the first-course block (wave radiation and propagation) is introduced by the postulation of a "generalized Coulomb's Law" (when the source charge moves and is subjected to acceleration) through the following steps:

1. Review of Coulomb's Law (electric field produced by a stationary charge).
2. Extension of Coulomb's Law to the electric and magnetic fields produced by a charge moving with uniform velocity much smaller than the "velocity of light" (non-relativistic limit), as summarized in Figure 9.1 [5]. Evidently, this

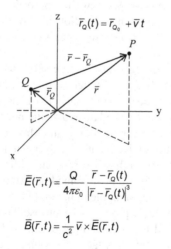

$$\bar{E}(\bar{r},t) = \frac{Q}{4\pi\varepsilon_0} \frac{\bar{r} - \bar{r}_Q(t)}{\left|\bar{r} - \bar{r}_Q(t)\right|^3}$$

$$\bar{B}(\bar{r},t) = \frac{1}{c^2} \bar{v} \times \bar{E}(\bar{r},t)$$

FIGURE 9.1   Fields produced by a moving charge (v ≪ c).

$$\tau = \text{retardation} = r_{QP}/c,$$

$$r_{QP}(t-\tau) = \left|\bar{r}(t) - \bar{r}_Q(t-\tau)\right| = r_{QP}(t) + \Delta r_{QP},$$

$$\frac{|\Delta r_{QP}|}{r_{QP}} \le \frac{|\Delta r_Q|}{r_{QP}} \le \frac{v\tau}{c\tau} = \frac{v}{c} \ll 1; \quad r_{QP}(t-\tau) \approx r_{QP}(t)$$

FIGURE 9.2   Approximations involved in Figure 9.1 for v ≪ c.

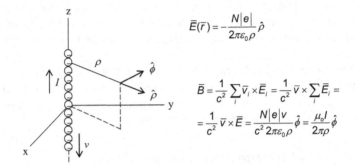

$$\bar{E}(\bar{r}) = -\frac{N|e|}{2\pi\varepsilon_0\rho}\hat{\rho}$$

$$\bar{B} = \frac{1}{c^2}\sum_i \bar{v}_i \times \bar{E}_i = \frac{1}{c^2}\bar{v} \times \sum_i \bar{E}_i =$$

$$= \frac{1}{c^2}\bar{v} \times \bar{E} = \frac{N|e|v}{c^2\,2\pi\varepsilon_0\rho}\hat{\phi} = \frac{\mu_0 I}{2\pi\rho}\hat{\phi}$$

FIGURE 9.3   Computation of the magnetic field produced by a filamentary current formed by moving electrons. N is the number of electrons per unit length.

procedure requires additional postulating that the electromagnetic perturbations propagate with the velocity of light, even though in the approximate expressions for the fields the retardation is neglected (v≪c; see Figure 9.2). Students accept this postulate very well if from the beginning the electromagnetic nature of light is recalled since the concepts of the "speed of light" and the measurement of distance to a star in light-years are usually very well established concepts in the student's mind.

At this point, it can be shown how this approach to the magnetic field meets with the standard presentation of the Biot and Savart or Ampère's Law, as studied in the freshman physics course, when filamentary conductors are considered (Figure 9.3).

3. Further extension to the case of a charge subjected to acceleration (also for v≪c) (refer to Figure 9.4) [6]. Evidently, this postulate includes as particular cases both step 2 (zero acceleration) and step 1, (zero acceleration and velocity). Therefore, this single postulate avoids the need of several other separate postulates (Coulomb, Ampère, Faraday, and Maxwell).

The basic properties of the radiated fields (behavior with r-1, spatial orientation) are easily drawn from these expressions without need to assume sinusoidal time variation, as it is usually assumed in textbooks when using the vector potential approach to obtain dipole radiation [3,7,8].

4. Computation of the fields produced by an oscillating charge, which can be immediately extended to an oscillating dipole and to a current element, as summarized in Figure 9.5, this latter by counting the number of oscillating charges within the element.

$$\bar{E} = \bar{E}_{stat} + \bar{E}_{rad} \quad , \quad \bar{H} = \bar{H}_{stat} + \bar{H}_{rad} \quad (\bar{E}_{stat}, \bar{H}_{stat}, \text{ fields for zero acceleration})$$

$$\bar{H}_{rad}(\bar{r},t) = \frac{Q}{4\pi c} \frac{1}{r} \bar{a}(t-\tau) \times \hat{r}$$

$$\bar{E}_{rad}(\bar{r},t) = \eta \, \bar{H}_{rad}(\bar{r},t) \times \hat{r} \qquad (v \ll c)$$

FIGURE 9.4 Fields produced by an accelerated charge located, at time t, at the origin of coordinates.

oscillating electron: $\bar{r}_e = \hat{z} \, z_0 \sin\omega t, \quad \bar{v}_e = d\bar{r}_e/dt = \hat{z} \, v_0 \cos\omega t$

oscillating dipole: $\bar{p} = -\hat{z}|e|z_0 \sin(\omega t) = \hat{z} \, p_0 \sin(\omega t)$

current element: $\bar{I}_0 \cos\omega t = -N_I |e| v_0 \cos\omega t \, \hat{z}$

$$\bar{H}_{rad}(\bar{r}) = j\frac{A}{r} e^{-jkr} \, \hat{z} \times \hat{r} = j\frac{A}{r} e^{-jkr} \sin\theta \, \hat{\phi}$$

$$\bar{E}_{rad}(\bar{r}) = j\eta\frac{A}{r} e^{-jkr} (\hat{z} \times \hat{r}) \times r = j\eta\frac{A}{r} e^{-jkr} \sin\theta \, \hat{\theta},$$

where $A = \dfrac{-|e| v_0}{2\lambda} = \dfrac{p_0 \omega}{2\lambda} = \dfrac{I_0 \, \Delta l}{2\lambda}$, depending on which case

FIGURE 9.5 Expressions for the fields produced by the oscillating elements quoted at the top.

5. *A formal difficulty: Power flux density.* This approach does not allow a formal introduction of Poynting vector, which has to be further postulated. However, a heuristic derivation is simple and, in the author's experience, satisfactory to students. The derivation runs in the following steps:
   a. Observation that the light power intercepted by flat surface is proportional to the surface area and to the area orientation, such that it can be expressed as the flux through the surface of a vector aligned with the wave "direction of propagation."
   b. Observation that the vector $\bar{S}(\bar{r},t) = \bar{E}(\bar{r},t) \times \bar{H}(\bar{r},t)$ keeps formal analogy with the expression $P(t) = V(t)I(t)$, has the expected direction, and has the proper dimensions of W/m².

At this point, this introduction to spherical waves has several advantages beyond the optimization of student effort over the conventional one that starts from the vector potential in sinusoidal time variation. In the first place, it sets an intuitive distinction between the far field and the near field, and the identification of this latter as the static situation produced by the same moving charges in the absence of acceleration. Second, it introduces students to topics generally considered "advanced," such as Rayleigh scattering [9] (the blue color of the sky [10], limiting attenuation in optical fibers [11]), Bremsstrahlung (X-rays) [12], and the fundamental instability of the Rutherford atomic model in the classical physics frame [13]. The cases of Rayleigh scattering and Bremsstrahlung are, besides, amenable to a simple quantitative analysis (exercise level).

9.2.2 PART II: PLANE WAVES

1. Plane waves are introduced as spherical waves seen from a distant, spatially bound region where both wavefront curvature and amplitude decrease with distance are negligible, as explained in Figure 9.6.
2. Reflection of waves at a flat conducting surface. Since in this approach the waves are connected to their source, this problem can be easily tackled with the help of the "method of images" for a flat conducting surface (Figure 9.7) [14]. Usually, this method is introduced in a basic course in electricity and magnetism and only needs revision here to check that it also guarantees proper boundary conditions at the conducting surface when radiation fields are considered. Introduction of the method at this point causes no problem to the course development.

    Note that Figure 9.7 can be interpreted both for the reflection of a spherical wave and, as a particular case, for a uniform plane wave.
3. Reflection/refraction process at a plane dielectric interface. In this case, the source approach to waves does not add any advantage, and the usual demonstration from continuity conditions at the dielectric interface is used, which brings the discussion to the presentation of this property without recourse to Maxwell's equations.

Even mathematically minded students find satisfactory an incomplete proof based on continuity of power flux density across the boundary:

$$\bar{E}_{\text{tan}}^{(1)}\left(\bar{r},t\right) \times \bar{H}_{\text{tan}}^{(1)}\left(\bar{r},t\right) = \bar{E}_{\text{tan}}^{(2)}\left(\bar{r},t\right) \times \bar{H}_{\text{tan}}^{(2)}\left(\bar{r},t\right)$$

(here, (1) and (2) refer to the dielectrics, and the subindex tan to the tangential components). Of course, this requirement means sufficiency for the continuity conditions, but not necessity. A further observation is that, in passing from one isotropic medium to another isotropic medium, no changes in the direction of either electric or

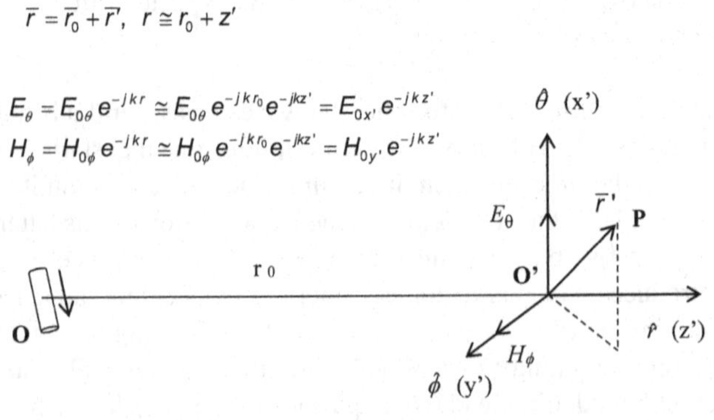

$$\bar{r} = \bar{r}_0 + \bar{r}', \quad r \cong r_0 + z'$$

$$E_\theta = E_{0\theta}\, e^{-jkr} \cong E_{0\theta}\, e^{-jkr_0} e^{-jkz'} = E_{0x'}\, e^{-jkz'}$$
$$H_\phi = H_{0\phi}\, e^{-jkr} \cong H_{0\phi}\, e^{-jkr_0} e^{-jkz'} = H_{0y'}\, e^{-jkz'}$$

FIGURE 9.6 A uniform plane wave as a spherical wave observed far away and attached to a new set of axes.

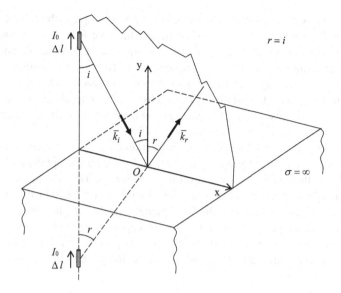

FIGURE 9.7 The problem of a wave (spherical, but also plane in the proper distant region) incident on a conducting plane as solved by the method of images. Here the case of a vertical dipole, which results in the electric field in the plane of incidence, is considered.

magnetic field should be expected (symmetry considerations). Therefore, only one pair of improbable, reciprocally inverse constants is left open to complete the proof.

9.2.3 PART III: OPTICS

At this point the core of the course in what concerns its presentation from the "generalized Coulomb-Ampère Law" has been presented. In the author's experience, it comfortably takes seven weeks out from the 14-week course duration. The rest of the course being presented does not depend on this approach, but it is described since it completes a fairly thorough introduction to electromagnetic waves that leaves the student in an optimum position to proceed with radio and optical communication topics [4]. Also, the full presentation of the syllabus will allow the reader to assess the pros and cons of this approach and how far one can proceed without recursion to Maxwell and differential equations.

The rest of the course devotes four weeks to an introduction to wave diffraction and optics and leaves the three final weeks for the preparation of a short project. The choice of optics as the final part of the course also enhances the possibility of simple lab experiments and projects concerning waves, given the availability of inexpensive lasers (either diode or He-Ne), detectors, lenses, mirrors, photographic and CCD cameras, diffraction gratings, etc.

1. The Huygens–Fresnel principle (expressions of the vector fields radiated from a small aperture illuminated by an incoming wave). According to the course philosophy, this principle is postulated, emphasizing its analogy with the fields

radiated by an elementary dipole. A first, basic example is the full electromagnetic description of Young's experiment, which, besides its clear approach to the phenomenon of wave interference, is amenable to simple lab experiments (either at microwave or optical frequencies) and is a good preliminary work for later study in arrays at the course in antennas and radio links. A second interesting example is that of the diffraction by a circular aperture, to be used in the next step and also with the advantages of preliminary work for later study in reflector antennas (circular reflecting dish).

2. Geometrical Optics. The previous diffraction concepts allow understanding of the introduction of ray approximation and geometrical optics, and steps B2 and B3 allow a quick arrival at mirrors and lenses. This section ends with the point spread function of an ideal lens or mirror (diffraction limit) which is introduced as the image in the focal plane of an object (the Airy pattern) produced by the aperture at infinity. This concept is of great interest to communication engineers since it permits understanding of the limiting performance of integrated circuit technology and of optical storage (CD-ROM, DVD).

9.3 COURSE PROJECTS

The main course advantage is, as already pointed out, its mathematically unobstructed way to electromagnetic waves that allow the student to concentrate on their physical properties and applications. In particular, there is a three-week period for the students, working in groups of three[1], to prepare a small laboratory-based project involving basic properties and inexpensive and easily available material. These projects allow the student to reach higher cognitive levels beyond the three first cognitive domains (knowledge, comprehension, and application), as defined by Bloom in his taxonomy [15]. Most courses and textbooks in electromagnetic theory have usually as implicit targets these first three domains.

To emphasize this shift from mathematics to physical meaning, the examples of projects are given below, some of which could hardly be approached in a conventional one-semester sophomore course on electromagnetic theory. The projects comprise six basic activities: Study of the phenomena involved, design of the experiments, lab work, discussion of results, report preparation, and oral presentation to their classmates.

9.3.1 Projects at Microwave Frequencies (10 GHz)

All these projects can be performed with inexpensive microwave components (Gunn diode oscillators, SWR or power meters, horn antennas).

1. Doppler shift measurements. This project can be performed with a Gunn diode oscillator, a receiving diode, and a pendulum with a metallic bob. Since the Doppler frequencies lie in the range of tens or hundreds Hz, the receiving diode output can be easily recorded with help of a PC sound card.

2. Young's and diffraction experiments.

3. Design and construction of a zone plate antenna [16]. With mechanical care, a "good dose of luck" and a commercial low noise amplifier (LNA) front-end, a 1.2 m wooden zone plate antenna can be used to receive satellite TV broadcasts.
4. Diffraction experiments with wire grids (diffraction gratings).

9.3.2 PROJECTS AT VISIBLE WAVELENGTHS

A laser source (a laser pointer is enough), a single-lens reflex (SLR) camera, polaroid filters, simple mechanical components (holders, rotators, clothespins), microscope glass slides, and adhesive tape is all that is needed to perform these projects.

1. Young's and diffraction experiments. Spectacular photographs in vivid red color can be obtained with help of a laser diode and a reflex camera with its lens removed so that the film acts as the projecting screen. The clearly visible speckle effects are also a very good source of discussion.
2. CD and DVD disks as diffraction gratings. Very accurate measurements (up to 1%) of track separation in a CD-ROM or DVD disk can be easily performed by using even a toy laser pointer and careful measurements of beam deviations upon a distant wall if the laser wavelength is properly calibrated. Also, pictures of the diffracted beam, obtained as in project 1, provide additional information about the number of diffracting lines intervening in the process.
3. Non-destructive identification of minerals and gems (with a flat surface). A fairly accurate measurement of Brewster angle, and, therefore, of refraction index, can be carried out with a laser pointer, two polaroid filters, and a rotating holder. Angles can again be measured by projection on a distant wall.
4. The polarization of skylight. The degree of polarization of light coming from different regions of the sky can be measured with a single-lens reflex, photographic camera fitted with a telephoto lens to cover a small angle of vision, and a rotating polaroid filter. The camera photometer provides a (relative) measurement of light intensity.
5. Construction of half and quarter-wave plates using adhesive tape. Very good results can be obtained by trial and error for laser light. When working with white light, beautiful chromatic results can be observed.

9.4 HOW FAR CAN THIS WAY LEAD TO? WHAT DO STUDENTS LOSE?

It has been explained in the introduction that this course is followed by other one-semester courses on guided wave and wireless communication systems, which include topics on antennas, transmission media (TEM lines, waveguides, and optical fibers), and microwave components. It is relevant, therefore, to ask whether the introduction of Maxwell equations is at some point indispensable.

Concerning the basic concepts on antennas, knowledge of the far fields produced by both a current element and an aperture are sufficient for the analysis of wire and aperture antennas if advanced topics -such as a formal derivation of the reciprocity theorem- are avoided. Concerning the transmission media and microwave components concepts, an early difficulty is the obtainment of propagation modes in waveguides and optical fibers, and of resonant modes in a cavity. With regard to the former, the modes of the rectangular waveguide can be introduced from the successive reflections of a plane wave between a pair of parallel conducting planes [17]. This approach allows the student to understand all the interesting characteristics of the "propagation mode" concept (cutoff frequency, dispersion, field structure, evanescence), and although the approach does not allow analysis of a circular waveguide, postulation of their modes does not suppose any break in logical rigor; optical fiber mode analysis is an advanced topic beyond the curriculum level. Similarly, resonant modes in a cavity can be presented as standing waves within a waveguide in a straightforward manner. In conclusion, Maxwell's equations can be omitted in the curriculum without loss of connection between the course and the subsequent courses dealing with communication systems.

But, of course, there is a loss. This paper does not claim an approach that is equivalent to the standard Maxwell's equations approach. The paper tries to find a reasonable alternative teaching strategy *given that* curricula transformation progressively reduces students' dedication to basic subjects – not only to electromagnetic theory, but also, for instance, to solid-state physics, photonics, and their quantum mechanics, and thermodynamics support. The question is how to manage the losses so that the graduate still has a sound foundation in basic scientific disciplines, yet he or she can cope with the ever-expanding technological advances.

In the case of this approach to electromagnetic theory, what is this loss? Concerning rigor, while it is true that the fields postulated for an accelerated charge are only valid for particles moving with velocities much smaller than the speed of light, this low-velocity approximation is the case for all engineering situations of interest, even for the fastest electrons in a semiconductor. Apart from this approximation, these fields contain all the physical phenomena predicted by Maxwell's equations. What the student does indeed lose is capability of analysis of boundary value situations, as found in the examples of mode propagation within a waveguide or an optical fiber.

The author's experience is that in the three-year program in which the course belongs (which, as pointed out in the introduction, for the purpose of this manuscript is comparable to a U.S. college bachelor's program), the evident losses introduced by the course approach are more than compensated by the more efficient use of students' time. Some of these students (a decreasing number, anyway) will certainly have to learn Maxwell's equations formally to handle boundary value problems, either because they proceed with the "senior engineer" degree after graduation or because of professional needs; but if care is taken to avoid rediscovering what is already known, this learning can certainly be easier and faster.

9.5 CONCLUSIONS

The author and colleagues have been using this approach to the course for the last four years, after three previous years of a standard approach based on the preliminary introduction of Maxwell's equations. They have found that with the new approach:

1. The student can be working with the vector description of electromagnetic waves and with relevant scientific and technical examples from the second course week, and his/her efforts are directed more to the use and interpretation of waves than to their mathematical derivation.

2. The course core (spherical and uniform plane waves, power flux-density, reflection, and refraction) comfortably takes seven weeks of the 14-week course, considering a weekly load of 2.5 hours of lectures, 1.5 tutored group work, and 6 additional hours of both students' personal and group work.

3. In the first half of this course, the student can already tackle problems of technical interest, which are usually considered advanced topics, such as the fields produced by a wire antenna, the radiation produced by a TV CRT display (Bremsstrahlung), or Rayleigh scattering in an optical fiber.

4. Probably, the most noticeable difference between this course approach and the standard Maxwell's equations one is the improved students' perception of the subject. Questionnaires show a change from the subject being considered one of the least interesting to one of the most interesting, as compared to the other sophomore course subjects (electronics, computer programming, circuits and linear systems, and signal processing). Still, the students consider the subject as one of the more difficult.

5. As a counterpart, students lack the ability to solve boundary value problems, and those who want to proceed with advanced topics in transmission media or microwave communication systems will eventually have to go through Maxwell's equations.

NOTE

1   In fact, the whole course is approached in a cooperative learning strategy and the students study and work in groups of three all the time.

REFERENCES

1. "Criteria for accrediting engineering programs" (Effective for the 2001-2002 Accreditation Cicle), Accreditation Board for Engineering and Technology, Inc. (ABET). 111 Market Place, Suite 1050, Baltimore, MD 21202, p.1. Available at http://www.abet.org.

2. P. A. Tipler, *Physics for Scientists and Engineers*, 3rd Edition New York: Worth Publishers, Inc., 1991

3. S. Ramo, J. R. Whinnery and T. Van Duzer, *"Fields and Waves in Communication Electronics"*, New York: John Wiley & Sons, Inc., 1965, pp. 642–646.

4. J. Bará, *"Ondas electromagnéticas en comunicaciones"*, Barcelona: Edicions UPC, 2001 (in Spanish). Also available in electronic format at http://www.edicionsupc.es.

5. W. K. H. Panofsky and M. Philips, *"Classical Electricity and Magnetism"*, 2nd edition, Reading, Mass.: Addison-Wesley Publishing Company, fifth printing, 1972, p. 346.

6. Ibidem, p. 358.

7. J. D. Jackson, *"Classical Electrodynamics"*, New York: John Wiley & Sons, Inc., 1962, pp. 268–273.

8. M. F. Iskander, *"Electromagnetic Fields and Waves"*, Englewood Cliffs, New Jersey: Prentice-Hall, Inc., 1992, pp. 641–650.

9. K. D. Möller, *"Optics"*, Mill Valley, California: University Science Books, 1988, pp. 556–561.

10. E. Hecht and A. Zajak, *"Optics'*, Wilminton, Delaware: Addinson-Wesley Iberoamericana, 1986 (Spanish translation of the English 1974 edition), pp. 256–258.

11. J. Gowar, *"Optical Communication Systems"*, 2nd Edition, New York: Prentice-Hall, 1993, p. 93.

12. W. K. H. Panofsky and M. Philips, *"Classical Electricity and Magnetism"*, 2nd edition, Reading, Mass.: Addison-Wesley Publishing Company, fifth printing, 1972, pp. 361–363.

13. L. N. Cooper, *"An Introduction to the Meaning and Structure of Physics"*, New York: Harper & Row, 1968, pp. 455–456.

14. R. S. Elliot, *"Electromagnetics"*, New York: IEEE Press, 1993, pp. 142–148.

15. B. S. Bloom, *Taxonomy of Educational Objectives: Book I Cognitive Domain*, White Plains, NY: Longman, 1956.

16. K. D. Möller, *"Optics"*, Mill Valley, California: University Science Books, 1988, pp. 366–368.

17. S. Ramo, J. R. Whinnery and T. Van Duzer, *"Fields and Waves in Communication Electronics"*, New York: John Wiley & Sons, Inc., 1965, pp. 386–387.

10 Taking Electromagnetics Beyond Electrical and Electronics Engineering

Soo Yong Lim

Department of Electrical and Electronic Engineering,
University of Nottingham, Malaysia

Yana Salchak

Griffith University, Australia

Cynthia Furse

University of Utah, USA

CONTENTS

10.1 INTRODUCTION

The evolution of technology has resulted in many disciplines becoming closely inter-twined over the course of history. Thanks to the rapid advancement of Information and Communication Technology (ICT), the world is becoming smaller with the clear distinct lines guarding the boundary of various disciplines more blurred as days pass. Having the possibility to receive, manipulate, and transmit large amounts of data digitally has not only transformed most of the existing sectors of life but has also served as a development agent for the various fields of science. ICT advancement and CS solutions that enable powerful computational capacities, numerical analysis, and automated processing to be consolidated have allowed for classical knowledge to be applied in a novel way to solve complex real-world problems, which subsequently has resulted in new interdisciplinary fields. To name a few, robotics, nano- and bio-technology, including biomedical engineering and biolectromagnetics are some prime examples, they are a product of the integration of classical scientific knowledge with emerging technologies [1–3].

The rise of interdisciplinary areas is inevitable since many of the challenging global issues today cannot be solved based on the knowledge in one field. In order to give undergraduate students the capacity to integrate parts of the comprehensive knowledge and skills in a unique way, attempts to facilitate the change of educational systems by incorporating interdisciplinary frameworks were being taken [4–7]. IT and CS undergraduate students gain a valuable competency in using highly demanded technological tools by the completion of their degree. However, deeper knowledge in related disciplines can be advantageous in the further development of their careers. EM can be considered as one of the most appropriate and promising disciplines for integration within their curriculum for this purpose.

Conventionally, EM is listed as at least one, and in some universities, two, required courses in the curriculum of electrical and/or electronic engineering. In addition, EM may be taken as an elective course, in the form such as optics, advanced EM, microwave, antenna, EMC/EMI, RF, microwave engineering, etc. For the CS and IT undergraduates, however, EM is traditionally listed as neither core nor elective. In this chapter, we discuss the practice of introducing an elective EM course to CS and IT curriculum, considering the specifics when teaching non-EE majors, who do not have a strong theoretical background in that area.

The proposed approach and the structure of the elective EM course were well received by the students, with many giving positive feedback. As the focus was to engage students by providing them with a visual representation of the concepts and engaging them with real-life examples, a similar approach could be beneficial when introducing EM to a general broader audience of researchers, students, and practitioners who are working in the related fields. This chapter also discusses biolectromagnetics in particular, as one of the interdisciplinary fields that require an intuitive understanding of EM fields and the basic physics principles behind its interactions with biological systems.

10.2   EM AS AN APPETIZER COURSE FOR CS AND IT STUDENTS

An appetizer EM course was added to the elective courses of CS and IT curriculum [8]. It was envisioned during the creation of this course that upon course completion, students would be able to better comprehend the physics behind the system they work with, even if they do not eventually pursue a career specializing in EM.

10.2.1   INCLUSION OF PRACTICAL LABORATORY

At a glance, the course content of this EM as an appetizer course has been designed to not fully cover mathematical theories and derivation but rather to give students practical experience on how to use these concepts to solve real-life technical problems. To that ends two practical lab sessions have been included as part of the course structure. On one hand, Lab 1 presented the visualization of waves to students, and that began from a very fundamental concept such as how frequency and time-domain signals are related. On the other hand, Lab 2 gave them a hands-on experience to transmitting and receiving RF signal using dipole antennas, RF cables, RF signal generator, and spectrum analyzer at three different frequencies, namely, 900 MHz, 2.4, and 5.8 GHz. As it turned out, CS and IT students appreciated and enjoyed the lab sessions even more than an engineering student would. A standard curriculum of CS and IT focuses primarily on how computers work, with courses offered to revolve around various languages of programming, algorithms, databases, software, and the like. As such, students do not get to do extensive hands-on experiments with equipment. Hence, when practical lab sessions were introduced to students in this appetizer EM course, they had a fun time assembling various dipoles antennas to the RF signal generator using RF cables, and visualized the received signal strength as displayed on the spectrum analyzer. This activity has sparked interest of students for EM, and along the way it has also deepened their appreciation for human–computer interaction, which is in line with the objectives of the degree they pursued.

10.2.2   A CREATIVE APPROACH TO CONDUCTING THE COURSE

Since CS and IT students have non-EE backgrounds, a creative approach was adopted to conduct this appetizer EM course. Its content revolves around topics that are relevant to modern life, which students can relate to. As an example, this appetizer course began with something interesting and relevant aimed at capturing students' attention, by engaging them with a device dear to them – the smartphone. An average youngster checks their smartphones uncountable times in a day. When this course prompted them to check the received signal strength on their mobile phones, to find out what is the signal strength they receive from their mobile network operator, it was a natural and easy way to spark their interest. Depending on which smartphone they were using, they can go to Settings > About Phone/Device > Status > Signal Strength to look at the received signal strength in one particular location. In [9,10], the readers can learn how to visualize the signal strength in current operating system (OS), for instance, iOS 14. Straightaway students were thrilled to move around the classroom. And in so doing, they observed that the signal strength may fluctuate, i.e., it might decrease from –80 dBm to –85 dBm, and go back up again. During this activity, students discovered that the signal strength provided by various network operators, i.e., that of Celcom, Maxis, Digi, U-Mobile, etc., in Malaysia, differ, in a particular location.

At this point, students were keen to find out more on this topic, and we proceeded to introduce them to power measurement basics and the advantages of using the decibel (dB). Students were excited to see that with logarithmic unit, the mode of power gain expression, covers a wide range of ratios with a minimal span in figures. They also learned that multiplication of numeric gain is replaced by addition of dB magnitude. A good knowledge of the decibel will come in handy to CS and IT students, as it is used for a wide variety of measurements in science and engineering, i.e., in acoustics, electronics, control theory, etc. Below are two sample equations that students will encounter. These were used for some tutorial exercises, as an introduction to decibel. Also imparted to the students are the mathematics of dB scale and their

TABLE 10.1
Mathematics of Decibel Scale

Single Unit	Multiple Units	Implication
dB + dB	dB	Product of two numbers
dB – dB	dB	Ratio of two numbers
dBm + dBm	Nil	Product of two power levels (not allowed)
dBm – dBm	dB	Comparison of two power levels
dBm + dB	dBm	Amplification of power
dBm – dB	dBm	Attenuation of power

implication, as summarized and presented in Table 10.1. Further detail about this appetizer course is available from [8].

$$\text{Gain}\left(\text{dB}\right) = 10\log_{10}\left(P_2/P_1\right) \tag{10.1}$$

$$\text{Power}\left(\text{dBm}\right) = 10\log_{10}\left(P/1\,\text{mW}\right) \tag{10.2}$$

10.2.3 IMPACTS ON STUDENTS

This appetizer EM course was class-tested among CS and IT students and it was well received. Many have found the appetizer EM course useful even though they were not EE students and had limited understanding of EM prior to taking the course. In retrospect, we observed that this appetizer EM course has left a positive and constructive impact on the students. One student commented that the course has greatly expanded on his own understanding of what an EM wave represents and that he has gained a more solid insight into the meaning behind the equations he used, such as that of dBm that he saw on his mobile phone often. Another student commented that as a student she has experienced first-hand that after taking this appetizer EM course, her understanding and interest in EM has vastly increased in a short period of time. Still some other students had this to say, that the appetizer EM course has helped them to expand on their future career choices, so that they not only were looking at the possibility of becoming programmers or computer analysts upon graduation, but also network/system engineers and researchers, or the like.

10.3 EXAMPLE OF EM IN A NON-EE DISCIPLINE: BIOELECTROMAGNETICS

Bioelectromagnetics (how fields interact with the body) combines biology, medicine, and electromagnetics. As multidisciplinary as this topic is, it is often beneficial to explain electromagnetics in ways that can be understood by researchers and practitioners in biology and medicine, as done in [11]. The audience for this book is medical technicians who use equipment that uses EM principles such as medical resonance imaging (MRI), impedance spectroscopy, radio-frequency cardiac ablation probes (RFCA), and other types of coaxial heating systems, hyperthermia systems, etc. Bioelectromagnetics is explained through graphics, such as shown in Figure 10.1,

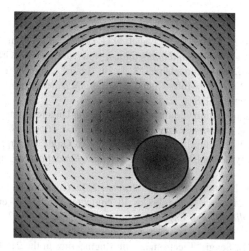

FIGURE 10.1   Calculated **E** fields in a two-dimensional model of a saline solution with a conductivity of 0.6 S/m, to which a round object with conductivity of 8 S/m has been added, in a round non-conducting dish exposed to a uniform 60 Hz **B** field in a direction out of the paper. The conducting object has lower fields and perturbs the field pattern both inside and outside of itself. (From [11])

and algebra (no calculus). This method of teaching a fundamental, intuitive understanding of electromagnetics can also be helpful for engineering students with an EM emphasis. Most EM courses try to help students understand the fields through mathematics. Enhancing this explanation with visualization and basic intuitive concepts can significantly improve students' understanding of electromagnetics. This type of teaching can be applied equally well for both EM-focused students and non-EM focused students.

10.4   CONCLUSION

Overall, we believe that taking EM beyond EE is a worthwhile effort. Undergraduates in IT and CS disciplines gain extremely useful and cutting-edge technological skills, which could be applied both in academic and non-academic sectors. Creating a practice of teaching EM that would not require *a priori* knowledge and would have a strong visual context and real-life examples to support the explanations, could make a huge impact on how this field is perceived making it easier to enter for non-EM majors. In the rise of interdisciplinary fields of research, acquiring sufficient knowledge of basic theoretic concepts and understanding of how EM fields work can be extremely advantageous as it explains the fundamentals behind the most of today's technologies.

ACKNOWLEDGMENT

The authors would like to thank Dr. Hugo Espinosa for his professional review and constructive comments on our chapter.

REFERENCES

1. M. P. Groover, M. Weiss, R. N. Nagel, N.G. Odrey, and A. Dutt, *Industrial robotics: technology, programming, and applications*, New York, NY: McGraw-Hill, 2012.

2. H. R. Jamali, G. Azadi-Ahmadabadi, S. Asadi, "Interdisciplinary relations of converging technologies: Nano–Bio–Info–Cogno (NBIC)", *Scientometrics*, vol. 116, no. 2, pp. 1055–1073, August 2018.

3. G. Rein, "Bioinformation within the biofield: beyond bioelectromagnetics", *The Journal of Alternative & Complementary Medicine*, vol. 1, no. 10, pp. 59–68, February 2004.

4. B. Tripp, and E. E. Shortlidge, "A Framework to Guide Undergraduate Education in Interdisciplinary Science", *CBE – Life Sciences Education*, vol. 18, no. 2, es. 3, May 2019.

5. M. Berenguel, F. Rodríguez, J. C. Moreno, J. L. Guzmán, R. González, "Tools and methodologies for teaching robotics in computer science & engineering studies", *Computer Applications in Engineering Education*, vol. 24, no. 2, pp. 202–214, March 2016.

6. L. B. Felsen, L. Sevgi. "Electromagnetic Engineering in the 21st Century: Challenges and Perspectives", *Turkish Journal of Electrical Engineering & Computer Sciences*, vol. 10, no. 2, pp. 131–146, May 2002.

7. Z.O. Abu-Faraj, "Bioengineering/biomedical engineering education and career development: literature review, definitions, and constructive recommendations", *International Journal of Engineering Education*, vol. 24, no. 5, p. 990, January 2008.

8. S. Y. Lim, "Education for electromagnetics: introducing electromagnetics as an appetizer course for computer science and IT undergraduates," *IEEE Antennas and Propagation Magazine*, vol. 56, no. 5, pp. 216–222, October 2014.

9. Juli Clover, "iOS 14 Includes Redesigned Field Test Mode," MacRumors, July 27, 2020 [text file]. Available: https://www.macrumors.com/2020/07/27/ios-14-redesigned-field-test-mode/ [Accessed: Nov. 26, 2020].

10. Jason Cross, "How to see your true cellular signal strength with the iPhone Field Test app," MacRumors, Mar 11, 2019 [text and video files]. Available: https://www.macworld.com/article/3346027/how-to-see-your-true-cellular-signal-strength-with-the-iphone-field-test-app.html [Accessed: Nov. 26, 2020]

11. J. Nagel, C. Furse, D. Christiansen, and C. Durney, *Introduction to Bioelectromagnetics*, 3rd Edition, Boca Raton, FL: CRC Press, 2018 (also translated into Korean).

11 HyFlex Flipping
Combining In-Person and On-Line Teaching for the Flexible Generation

Cynthia Furse and Donna Ziegenfuss
University of Utah, USA

CONTENTS

11.1 INTRODUCTION

Changing student populations, evolving learning contexts, and a shift in higher education priorities demand that faculty rethink the traditional face-to-face classroom teaching approach. Although 2020–2021 Covid constraints and challenges on campuses around the world placed restrictions on the classroom context, this situation was also the beginning for new opportunities and innovation. Flexibility, resilience, and creativity, as well as building on past success, can help faculty move forward to imagine the classrooms of the future. Dr. L. Dee Fink, an expert in higher education course design and faculty development, contends that "Faculty [lack of] knowledge about course design is the most significant bottleneck to better teaching and learning in higher education" ([1] p. 24). He is a proponent of "backward design" where faculty first articulate what they want students to know, be able do, and "be" at the end of instruction, and then work backward to align the assessments, and teaching activities to those goals [24]. Explicit instructional design work, curriculum revisions, and re-centering our teaching on the student can make a difference in learning success. HyFlex Flipping is one example of how a new teaching approach can integrate the learning that happens inside the classroom and outside the classroom. It provides flexibility to meet the variety of student needs and learning styles, creates a situational context for active engagement in learning, and provides opportunities for connecting classroom learning with real-world authentic learning.

Educational research on how people learn has led to some significant and very important findings about effective strategies for teaching and learning [2]. Students prefer to learn in different ways and have different learning styles [3–5], and many exhibit preferences for different learning modalities [6,7]. Although learning styles and learning environment modalities can impact the success of the student learning experiences, using active learning strategies, and designing the learning in a way to engage the students in the learning process has been cited in the literature as the most effective way to improve student success [8,9]. Active learning is a strategy that can be implemented and adapted in a variety of learning environment modalities, which makes it an emerging and commonly used strategy for STEM teaching [10]. Active learning in the face-to-face classroom could involve learning by "doing" group projects, such as problem-solving sessions, case study analysis, collaborative learning, or just peer to peer interaction. Online active learning can replicate the face-to-face activities utilizing technology tools and techniques. A third and emerging option for active learning is the HyFlex learning environment where the course is designed so that students can learn in the face-to-face, or online, or a combination of both modalities in the course.

From these findings, several strong teaching methods have emerged. Active Learning, which actively engages students in problem-solving rather than just passively listening to a lecture, is the new norm, to the point that not using it may be considered unethical [11]. Several methods now apply active learning as the key component of the class. Peer Instruction (PI) uses pre-class readings or short video lectures and poses questions to be solved in small groups during in-class time [12–14]. Problem-Based Learning (PBL) focuses learning around a real-world problem [15]. Just In time (JIT) lectures provide small instructional snippets as need arises during problem-solving [16]. Supplemental Instruction (SI) provides additional time for problem-solving, often with the guidance of a TA [17]. The Flipped/inverted/ hybrid class uses pre-class video lectures and/or readings to free up in-class time for active learning [18]–[24]. All of these methods of enabling active learning have been shown to increase student learning, retention of learning, satisfaction, and in some cases retention within the program [25–27]. But not every class using every method has been equally successful, and there is significant variability from instructor to instructor [18–24]. Successful classes, using each of these methods, have key features in common, features that address the key findings above: (1) Active learning engagement (which we will cover in section 11.4), (2) Multiple ways for students to learn (which we will cover throughout this chapter), (3) Alignment between the learning objectives and content and alignment with the real world (which we will cover in section 11.3) [28].

With the rise of technology, there are now many modes of course delivery – Face-to-Face (F2F, in-person), Remote Synchronous (RS, such as using Zoom), Online Asynchronous (OaS, such as a Massive Open Online Course (MOOC)), and Hybrid/ Blended combinations of these.

In this chapter, we will focus on a Hybrid-Flexible (HyFlex) model, that combines in-person (F2F), remote synchronous (RS), and online asynchronous (OaS) modes, enabling the students to choose and move between the modes as their lives and learning preferences dictate [29]. We build this design based on the Flipped Class, where recorded lectures and written content is provided for asynchronous pre-class viewing, F2F or remote synchronous (RS) time is used for active problem-solving (and recorded for online asynchronous (OaS) review or viewing), and active learning extends to online asynchronous (OaS) activities, as well. Students can move seamlessly between all modes of delivery. They may choose to join the F2F classroom one day, the RS class another day, and the OaS class a third day, of they may combine the activities, as their preferences and learning needs dictate. We choose this model, because it offers the greatest flexibility for both learners and teachers in times of uncertainty and change, as well as the promise of a truly flexible learning environment for now and for the future, where the life-changing opportunities of education can be provided anywhere, anytime, to anyone.

This chapter provides background and specific recommendations for creating a HyFlex class. We expand on our personal experience with the flipped classroom and active learning [18–24], [30,31] and helping other faculty develop their own flipped classes [18,32]. Section 11.2 covers how to plan your HyFlex class using Backwards Design – deciding what you want to teach (objectives), how you will know the students have learned the material (summative assessment), what content is needed and how to deliver it, how to provide meaningful practice (active learning) and immediate feedback (formative assessment). We will consider differences in learning styles, and the connection and communication strategies that are needed in all classes but that require extra planning for remote and online modalities. In Section 11.3 we discuss creating video content (lectures, examples, and establishing a teaching presence), including their type, content and length, and technology and tips for creating them yourself. We will also discuss best practices for teaching using material created by others (co-flipping [19]). Section 11.4 is devoted to Active Learning, which is the very heart of teaching and the key to success in all modalities. We will describe several methods for engaging students in all modalities, and suggest ways to blend them in a HyFlex model. In Section 11.5, we will discuss summative (end of term) and formative (feedback along the way) assessment, and suggest tools and strategies for providing assessments that are as flexible and student-centered as the HyFlex class. Specific recommendations are given throughout each section. The suggestions and recommendations in this chapter are appropriate for all types of classes and modes of delivery, but we will focus on examples from electromagnetics for a HyFlex class. In full disclosure, we emphasize that we have not yet taught a course in full HyFlex mode, and that all of the technologies and support services described in this chapter may not be available to all instructors and all learners. We will discuss some methods to bridge this digital divide, with hopes that electromagnetics and wireless communication will gradually break down these barriers and enable digital connection for all of the world's learners. We hope our suggestions and the theory behind them lead you to think of new and creative ways to apply these ideas in your own teaching, and we encourage you to share these ideas broadly with others who seek to teach the magic of electromagnetics. Let's begin…

11.2 DESIGNING YOUR HYFLEX COURSE

In this section, we will describe the four key things to think about when designing a HyFlex course, based on a student-centered flipped instructional model delivered in a flexible combination of face-to-face (F2F), remote synchronous (RS), and online asynchronous (OaS) modes. The four key things are (1) Learning Outcomes, (2) Assessment, (3) Content & Delivery, and (4) Practice with Feedback.

Let's start with an example:

1. **Learning Outcome:** An instructor wants to teach students how to do "design a single stub matching network"
2. **Assessment:** There will be a problem on the final exam (summative assessment) where the student is asked to design a stub network for a particular load impedance.
3. **Content & Delivery:** The instructor wants to spend time in class working through examples/homework problems designing series and parallel stubs terminated by open and short circuits (problem-solving) and also discussing why one might choose one of these designs over another in a practical application (real-world context). There is not enough time in a 50-minute class to give a lecture describing single stub matching in detail, and do these four examples, let alone discuss their practical tradeoffs. And this instructor knows the students will learn more, ask more questions, and remember what they learn longer if they work through the four examples individually or in a group, perhaps a combination of both (active learning/peer teaching). Of course, the students could do these for homework, but the instructor knows many will initially struggle, some will gain permanent errors in their methods, and most won't think about the nuances of their designs (higher level thinking skills). So the instructor decides (prioritizes) the most valuable use of the 50-minute

student–instructor class contact time is working through the four stub examples and discussing which to use in a practical application. ***Video Lecture Content:*** Because the students need to understand the basic steps for designing the stub network, the instructor posts a video lecture (created personally or shared from another faculty member), which the students will watch before class. The lecture is short (3 or 4 segments, each under 5 minutes long) and to the point. To be sure they watch the video, there might be a very short and easy quiz after the video, or credit for notes taken and uploaded, or just the practical incentive that it is much easier to learn if you watch the lecture. Some students don't like to watch videos, so they might read the textbook instead, but the instructor knows that most students pick up new material more easily with a video explanation than they do with a textbook explanation. Both are provided.

4. **Practice with Feedback:**

   Active Problem-solving: During class contact time, the instructor could do the four stub examples on the blackboard (for a F2F class), or on a tablet (for an RS class connected by Zoom/ GoogleMeet/ WebEx/ Teams, etc.). For this HyFlex course, the instructor will use a tablet (a nice way to show and draw on a Smith Chart), projecting it on the screen for the F2F students and sharing it via Zoom/etc. for the RS students, and recording and posting it for later use by OaS students (and review by everyone). Even better, the instructor wants the students to actively engage with these examples, so assigns small group breakout sessions (active learning and peer instruction), and asks each group to solve one of the four examples. The instructor may guide this problem-solving session step-by-step, or may just turn the students loose and answer questions as they arise, depending on the level of the students (advanced students typically need less guidance than beginners), how much time each problem takes (guidance can speed things up), and how complicated the problems are. Each group can then report back on their design, either verbally with screen sharing, or by uploading a document (a cell phone camera shot of handwritten notes for the F2F class or a screen capture of an RS class). For OaS students, they may do the same thing individually, or in their own small groups that meet synchronously but at a time of their own choosing, different than the class time.

   Problem-Based Learning (Real-World Context / Higher Level Thinking): The instructor then summarizes each of the designs and discusses their tradeoffs. Several examples of real-world scenarios could be provided, so the students could discuss in what scenarios their design might be the best, and why. Alternatively, working out matches for each of these scenarios could be a homework assignment, with discussion on which were best and why the following class period.

   Homework: For homework, the instructor assigns the students to do at least one more example (possibly for a real-world scenario, as described above), and check or simulate their design (feedback=formative assessment). The instructor might post a stub matching exam question from a previous year (either as homework or a low-stakes quiz), and ask the students to solve it in a specific amount of time, so they know they can do this efficiently and accurately (feedback). Students might also be asked to write a tutorial on how to design a single stub matching network, and share it with others for feedback (reflection, peer instruction).

(Back to 2) Assessment: Students may later build this in the lab (real-world context, summative assessment), and correctly answer an exam problem (summative assessment).

(Finish with 1) Outcome: How did they do?

11.2.1 BACKWARDS COURSE DESIGN FOR THE HYFLEX COURSE

When designing HyFlex courses, the prioritization of what will be done during student–instructor contact time, and the integration of the pre-class video lectures and the in-class active learning activities requires extra planning and reflection. Problem-solving that ties the pre-class videos to real-world applications (problem-based learning) is particularly engaging [15]. Backward class design helps instructors define what outcomes they seek, how they will assess those outcomes, what activities are needed to teach those outcomes, and where students will receive active practice and feedback for those outcomes. Planning where in this process the student–instructor and student–student contact time is most valuable is key to planning the HyFlex classroom.

Backwards course design for the HyFlex class starts planning with outcomes [1,33,34]:

1. *Learning Outcomes:* Define what you want students to be able to DO at the end of each learning session ("Be able to design widget x"). These are learning outcomes. Tell the students what you want them to learn, how you will assess this, and how each activity and assignment leads them toward that outcome. Tell them how this relates to the real world. Defining learning outcomes for the course, each section of the course (each chapter?), and each lecture will help guide both you and the students along the way.

2. *Assessment:* Define how you want students to be able to demonstrate they have mastered the outcomes (exams, homework, quizzes, projects, reports, labs, presentations, posters, etc.). Think creatively here, does your assessment really measure what you want your students to be able to DO? Could there be multiple ways a student might demonstrate proficiency in a topic (consider providing multiple, flexible assessments)? Can students have another chance(s) to re-learn and demonstrate proficiency if they mess up? Also remember to plan in both student learning assessments and assessments of how the course is going. We will discuss this further in section 11.4.

3. *Content & Delivery:* Define how you will teach the topic (lecture, demo, written material/textbook, etc.). Plan on at least two sources of information for every topic, and the more, generally the better (as long as they all agree). Think about what material students can best receive by listening (recorded lectures or

short lectures in-person or remote) or reading (textbook or supplementary materials), what material they are likely to struggle with (this is best for you to help them with during class time (in person or remote)), what material they will need more practice with (homework, projects, labs, etc.), and what prerequisite knowledge and skills are needed.

4. *Practice and Feedback:* Define how students will practice each outcome, and how they will get feedback (immediate, if possible). Feedback can be quizzes, working together in class/online, feedback from self, peer, TA, instructor, posted solutions. Multiple opportunities for practice and feedback, spaced over time, helps learning and retention. In addition to feedback provided to the students, what feedback will help you improve your course?

11.2.2 STUDENT-CENTERED LEARNING

Backward course design that focuses on the learning outcomes first results in a student-centered learning approach [35,36]. Thinking about what the student can do (outcomes), and how they and you will know it (formative and summative assessment) puts the emphasis on learning rather than on teaching. Integrating the outcomes, assessments, content, and practice/feedback provides a well-integrated teaching approach. Communicating this to the students guides a well-integrated learning platform. Paying attention to the prerequisite knowledge (and providing links to this information) also helps meet students where they are, and will help them learn better. Giving students immediate feedback will help improve their learning (active learning naturally provides some of this feedback). Considering individual learning styles [3] can create a very flexible, student-centered approach for which the HyFlex classroom is particularly effective. For example, for Global Learners [3], explicitly telling them what they are expected to learn that day, and how it is used in a real-world application, will help them understand it better. Consider include a sidebar about learning styles

11.3 CONTENT AND DELIVERY

There are many different ways to deliver lectures in a HyFlex model. The goal is to provide similar lecture support for both face-to-face and remote students (synchronous and asynchronous), without requiring too much extra time from the instructor. Think about what is needed in your lecture – the ability to write equations? Images? Demos? Other things? These are what both F2F and remote students need to be able to SEE and HEAR. So think about how to produce the best in person, video, and audio quality for both sets of students. One way to do this is to pre-record lectures, and have students watch those before class time. Another is to do short lecture segments live during class, and capture this live for students joining remote synchronously, and record it for the asynchronous students. (Capture technology of this sort is going to be important even if you are doing problem-solving and not lecturing in class, so any HyFlex instructor needs to consider how to do this.)

11.3.1 The Live HyFlex Classroom

Synchronous teaching time in the HyFlex classroom involves both F2F and RS students. This requires technology to make both live video and audio available to students in multiple locations. This is true, whether you are teaching a lecture segment or a problem-solving experience. We will start with what does not generally work well – lecturing for an hour on the blackboard, capturing this with a video camera from the back of the room, and streaming that live for remote students does not work well for a variety of reasons. First, it is extremely difficult for any student to pay rapt attention to any lecture for a full hour (estimates for adult attention span are 10-20 minutes), let alone a remote broadcast lecture, let alone when filmed in poor visual quality from the back of the classroom. Better approaches to lecturing for an hour is to either flip the classroom (short video lecture watched before class, with active problem-solving in class) or segment the lecture into several smaller portions, and have the students participate actively in between each segment. Both of these methods support the active learning that is so key to effective teaching.

Technology for sharing visual and audio information with both in person and remote students is very similar to that for producing video lectures, described below. A tablet PC, iPad, or paper+document camera allow the instructor to write and share information, while capturing this by video to share with the remote students (often by Zoom, etc.). Meanwhile, this same information can be projected onto a screen in the front of the class for the F2F students, or to their own laptops in front of them, and can be recorded for OaS students, and for review by anyone later. The instructor needs to remember that they can't personally be seen in this format, so pointing at the screen in front of the classroom to emphasize some aspect of a derivation will be invisible to the remote students. Also, the instructor needs to wear a wireless microphone in order to maintain audio quality as they move around and interact with the F2F students. The built-in microphone on tablets and laptops is not sufficient for HyFlex teaching. Also, F2F student questions or comments are not heard by the remote students, so the instructor needs to be in the habit of repeating this information so it can be heard. Gathering comments and questions from the remote students is also important. Regularly checking chat, or using polls (Google Forms, Kahoot, Zoom, etc.) can help gather these questions.

Alternatively, the instructor can write on a whiteboard (which generally shows up in video better than a blackboard) for the F2F students, and a videographer can capture and stream video for the remote students. This may be a little easier for instructors who are used to teaching this way, but it requires very good video capture/streaming. Some universities are now putting classrooms together to facilitate this, so in the future, this option may be commonly available.

One of the advantages of the technology-enabled HyFlex classroom is that if the instructor and all students have tablets of some kind, anyone can "point at" or annotate the screen, which enables asking and answering questions. Also, individuals or groups of students can share their work. For instance, after a problem-solving breakout session, each breakout group could share their work.

11.3.2 ONLINE TEACHING MATERIALS (VIDEO LECTURES)

For the flipped HyFlex class, short video lectures can replace in-class lectures, focusing live class time (F2F or RS) on active learning. The technology for creating these videos is what most faculty worry about, but best-practices for content and delivery are equally important, as described in this section.

Electromagnetics has used computer visualization extensively to make other-wise invisible fields visible [37]. Many video lectures are available, and the IEEE Antennas and Propagation Society recently established an online Resource Center for sharing of educational material, including full courses with video lectures. These include Basic Electromagnetics [30,38–41], Design of Electromagnetic Devices [42], Advanced Electromagnetics [10–12], Design of Electromagnetic Devices [42], and Numerical Electromagnetics [18,31]. Video content is also available on YouTube, MIT's OpenCourseWare [43], the Khan Academy [44], Udacity [45], Coursera [46], edX [47], Canvas Network [48].

11.3.2.1 Best Practices for Video Lectures

Video lectures provide learning content for all students in the HyFlex class (F2F, RS, OaS), free up time for active learning, and can be repeated and rewatched (one of the most popular aspects of this modality). Video lectures slightly outperform traditional in-person lectures for learning [49], and Millennial and Generation Z students strongly prefer them [50]. Video lectures should be short (15–20 minutes total for a traditional 50-minute class), and recorded/uploaded in even shorter chunks (3–5 minutes each). This helps students gather their thoughts and take notes in between, and they are easier for instructors to make without stumbling (thus eliminating the need to edit).

Video lectures are flexible – students can watch them when and how they wish to, repeat sections, speed up or slow down videos (such as for English as second language (ESL) students). Closed captioning is popular with many students who may choose to watch them on the bus or other noisy place, and they are important for hearing-impaired students. Closed captioning is available automatically through many video hosting services including YouTube. Adding something interactive within or between video segments also improves learning [51,52]. Short quizzes, problem-solving, predictions, and raising interesting questions can make the video more interactive.

11.3.2.2 Creating Video Lectures

There are numerous types of video lectures (see Module 2 of [18]), which should be chosen based on the content. For lectures that include math and derivations, PCs/tablets or paper + document camera [53] (a cell phone can serve as a document camera [54]) can project writing on the computer screen. Screen capture recording software (ActivePresenter, Zoom, etc.) records what is on the screen, either with or without also showing the instructor talking. Videos of an instructor writing on a whiteboard (blackboards tend to be difficult to see) can also be effective if the videography is well done. Audio quality is critical and is greatly improved by wearing a microphone (a wireless mic is handy). The video and audio quality of the lectures needs to be good enough that a student can view them easily even on a small device such as a tablet or cell phone (which is one of the reasons that full-blackboard videos are not as popular). But they don't have to be perfect. Small mistakes are inevitable (give students extra credit for finding them), and students like to feel connected to their very human instructor (so when your pet jumps into view, it's ok as long as it's not a distraction for too long). For HyFlex courses were students are RS or OaS, a tablet, phone, or document camera is a must for the students as well as for the professors, so they can share their work and questions with you.

Teacher presence (sharing the real person) is an important consideration, particularly for OaS students who may never see you in person. There are several good ways to do this, including making a short video clip of you giving an overview to the day (outcome, real-world context, learning activities, etc.), using a split-screen recording software (Zoom, Kaltura, ActivePresenter) that includes video of you talking along with the equation drawing/writing, including your photo on the front page of slides, and reaching out personally to students.

11.3.2.3   Hosting Video Lectures & Consideration of the Digital Divide

Videos can be hosted locally (most learning management systems now support video) or on sites such as YouTube and then organized and linked on the course website or learning management system. These can be made private or public. Public videos are often good publicity for the professor, department, and university and can have high impact for the electromagnetics community (consider contributing your videos and class to the IEEEAPS.org Resource Center). In areas where students do not have access to the Internet, alternatives must be considered. These include providing materials on DVD (if students have computers at home), school or library computers, or using reading materials rather than video materials for class preparation. A caution is provided here, however, that few students pick up as much material from reading as they do from video content. Many students use their cell phones or small tablets to watch the videos, mainly for convenience. Thus, it is important to keep the writing on the screen relatively large and legible, even if this means using more screens to complete a topic. It can also be helpful to provide screenshots or the original power point (with the written notes saved), so a student may print these out and take notes on them throughout the video.

11.3.2.4   Getting Past Video Creation Challenges

Creating video lectures is one of the biggest concerns instructors face with HyFlex teaching. Cost, time, and managing the technology are some of the concerns. Cost can be minimized, as inexpensive tablets, document cameras, and cell phones can readily be used (see Module 2 in [18]). For instructors who prefer the simplest possible technology, "hands writing" with a document camera or cell phone is generally the simplest approach. Audio quality should not be neglected, so generally it is beneficial to use a headset microphone when video recording. Cell phones are notable exceptions to this, as most of today's phones have excellent video and audio quality. Time to produce the videos is a concern the first time through a class, but most videos can be reused in subsequent semesters. Videos can also be borrowed from/shared with others (called co-flipping [19], just be sure to watch them, so you can teach the same way as the video!).

For a typical 50-minute class, the lecture should be about 15–20 minutes, broken into 3–5 minute chunks. Most instructors spend 4+ hours to record their first lecture video. With practice, they are much faster (about 1 hour). Avoid editing, because it takes so much time. Keeping the video segments short (3-5 minutes) helps, because if you mess up one segment, you can just start that segment over again without losing much. After the first semester, when videos are being reused, most faculty find the overall preparation time for a HyFlex class to be significantly less than for a traditional lecture class [19].

First-time video instructors can have personal concerns, as well. Some really want to be the "sage on the stage," though they quickly realize the advantages of choosing in-class time to maximize their importance to their students. Most people don't really like seeing and hearing themselves on camera. Many find it hard to talk to an empty board. Try starting with a few very low-risk high-reward videos such as doing examples or exam solutions, which students virtually always appreciate. A trick for lecturing to an empty screen is to provide a "face" to speak to. A small paper cutout of a familiar, jolly face (such as a comic character or your pet), taped so that it is peeking over the top of a laptop, may help create more animated lectures than talking to the laptop alone.

Copyright is important to consider, especially if you want to post your video publicly. Most universities have copyright librarians who can advise on this issue, and many textbook publishers are generous about giving permission to include figures, etc., in video lectures.

11.3.2.5 Getting Students to Watch the Videos

Getting students to watch the videos before class is essential, so they have the technical content for that day's active learning activities. The most important thing is to be sure the content of the videos is actually valuable to them. Video lectures should be quick (not rambling), to the point (be clear about what that point is), and at the right level. Then don't plan to repeat them during class (give a short recap, but don't repeat the lecture). Expect students to bring their notes to class and use them for the active problem-solving time. Typical watch-rates are about 80%. This can be assessed or at least estimated and perhaps incentivized through interactive activities in the videos themselves (short answers, short quizzes, short problems). As with any class, live attendance in the HyFlex class, depends on if class time is inherently valuable to the students. Thus, repeating the lecture, doing problems that are too easy or not related to upcoming exams will decrease classroom attendance. Like video watching, classroom attendance can be incentivized by having short quizzes or questions (classroom response systems have been used for this for many years) that basically mandate classroom attendance. However, (what not to do) effectively mandating that students attend class, but then not having the class be of sufficient value to them, will quickly alienate the students and make the class ineffective. Also, if the goal of the HyFlex environment is flexibility, mandating attendance at a particular time is antithetical to this goal.

It is not yet clear in the literature if it is better to use an inherent value model or a small incentive model to get students to both watch the videos and attend class. It is likely this depends on the student body, local culture, and culture set by the professor. Continually assessing video watching and class attendance throughout the entire semester is important for the HyFlex approach to be successful, so we recommend designing that assessment into your classroom approach, and talking to your students about the importance of both.

11.3.2.6 The Role of Readings

We have talked about video lectures, but what about readings? Can't students just read the chapter, and come to class prepared? There are definitely examples in disciplines such as law and business where this is the teaching norm, but this is more problematic in STEM disciplines, where individual readings may be more difficult to understand. This depends on the level of the students, their background, and the level and quality of explanation in the reading (some books are easier to read than others). Most students find watching a video easier than reading a book, and absorb the content more effectively in video format. As interactive e-books and interactive online learning platforms improve and reach electromagnetics, it is possible that readings of these types will match or replace video lectures. For now, we recommend providing video content, along with good quality written materials, thus enabling students to use both types of materials as they prefer. Even when video lectures are available, a small minority of students prefer to just read the chapter, and most use both.

11.4   ACTIVE STUDENT ENGAGEMENT

Active learning [13,55–59] – engaging students in actively applying course knowledge rather than passively listening – improves student engagement, learning, retention, confidence, and satisfaction with the class [60–63]. Active learning (also called engaged learning) is now the norm, either added to the traditional lecture, or displacing the in-class lecture entirely (as in the flipped class). This is arguably the most important aspect of the HyFlex classroom [60]. Active learning can be as simple as having students turn to their neighbor or join a Zoom breakout room to discuss a question, or as in-depth as group design projects. Active learning encompasses peer-assisted learning, problem-based learning, collaborative learning, cooperative learning, and peer tutoring, all of which, in and of themselves, have been shown to increase student learning in various ways [26]. Activities can take a minute or two, the whole class period, or an entire semester. Active learning is certainly not new to engineering, and electromagnetics in particular. Labs are a regular part of most engineering curriculum, and use of numerical simulations has been a mainstay of electromagnetics education for decades. Excellent resources for active learning strategies can be found in [57–59]. As starting suggestions, here are just a few simple and effective active learning strategies that are very useful in both F2F and RS classes, and suggestions for ways to adapt them for OaS:

1. *Think-Pair-Share*

    Think-Pair-Share (a form of peer instruction) or just Pair-Share [3,52] is a quick and easy way to engage students in class. THINK – Students think about a problem on her/his own; PAIR – Students turn to their nearest neighbors, or a breakout room or discussion board, and share their ideas; and SHARE – Student shares her/his ideas with a small group or the class. This is a great way of getting students to ask questions. Instead of asking, "Are there any questions?" (to which the answer is inevitably no), instead ask "PAIR: Turn to your partner, and find the most confusing thing about today's topic." And then SHARE with the class, which inevitably generates good questions and discussion [64], Ch. 23. Think-Pair-Share is also great for short answers or longer problem-solving in class [55]. This can also be adapted for RS and OaS classes using polling or breakout rooms.

2. *Muddiest Points*

Muddiest points [65] are concepts that remain confusing for students even after the lecture/class time. Muddiest Point [66] assignments ("What is the most confusing point this week? Try to answer it.") can help the instructor understand what the students are struggling with, and it also help the students reflect on their understanding. Muddiest Points also make a good weekly Discussion Board Topic.

3. *Classroom response systems, Polling and Discussion Boards*

Clickers (from TurningPoint), TopHat, LearningCatalytics, Kahoot, Polling (such as in Zoom), and other classroom response system (CRS) allow the instructor to pose a multiple-choice or short-text question, and each student to think about and answer that question. Some (Clickers) require special hardware. Others collect responses from mobile phone, tablet, or computer. These can be particularly nice for large classes, hybrid (F2F+RS) classes, for getting the class started, and for encouraging and assessing class attendance.

11.5 STUDENT-CENTERED FORMATIVE ASSESSMENT

The HyFlex classroom's student-centered approach warrants a more student-centered assessment approach. Summative assessments such as exams can tell how well the students have learned the material by the end of the course. Formative assessment such as self-checks, in class active learning activities and peer discussion, and course feedback throughout the semester can be used to adjust and improve both teaching and learning in real time [67,68]. Reflecting on their learning can help students be more effective [26,27]. Formative feedback strategies also helps professors adjust and improve the course in real time [69] and improve the culture of the learning environment [70]. A simple and easy-to implement formative assessment strategy for a flipped freshman circuits course is described in [23].

Formative feedback can improve both teaching and learning. This is particularly important for a HyFlex class, where the relationship between the instructor and students and between the students themselves is more dynamic and interactive. Formative feedback can be collected throughout the course (such as at 3-week intervals), asking about whatever the instructor is trying to accomplish (such as improving the labs, getting students to ask questions, assessing workload, etc.). For example, asking "What is going well with this course? What could be improved? What can I do to help you learn better? What can you do to help you learn better?" helps gather valuable feedback for improving the course in real time. One common surprise to instructors is that students' struggles in class are very often seated in challenges with school-life balance [66]. And although students can learn to manage their time better [71], it is incumbent on instructors to appropriately manage the workload expectations for their classes, and the HyFlex class in particular. Instructors should design their HyFlex class so that it no more work than a traditional class, and monitor the amount of time students are spending through regular student feedback.

11.5.1 Exam Grading Strategy

Exams and other summative assessments are meant to show how well students understand the material, but there are challenges. What if a student does great on the final but failed all the midterms (global learners [3] may do just that)? What if a student gets sick and can't come to class to take the exam? What if they sleep in and miss it? A structure for exam grading that uses the final exam as make up exams for the midterms can provide greater flexibility and motivation for the students, and less frustration and less grading work for the professor (see syllabus for [55]). Midterm exams are given throughout the semester, and the final is broken into individual parts, each reflecting the content from one of the midterms. Students may take any or all parts of the final. Their grade for each part will be based on the best of their midterm score or their score for that part of the final. Students who have done well on the midterms may choose not to take the final. Students who did poorly on some of the midterms can make them up, demonstrating they know the material. Students are motivated to learn material they thought they understood but discovered they didn't, and this method cuts grading time for the final about in half, which TAs and instructors generally appreciate at the end of a busy semester.

It is always important to have grading assessment (exams, assignments, etc.) align with the objectives of the course. This applies to both content and to how the course functions. For instance, if a major feature of the course is peer instruction and learning to work collaboratively and cooperatively in teams, grading individuals on a curve can significantly disrupt this process. Consider setting an absolute grading scheme, rather than a curve.

11.5.2 Labs

Labs in principle are more challenging to implement in a HyFlex setting, but should follow the same principles. Make them a personal and engaging experience. Plan your labs like you would your course (See 11.2. above). Get Creative About Locations. Read more about hands-on labs in Chapter 12.

11.5.3 At Home / Online Labs

Some labs may be able to be moved to online/at home. Review whether specific test kits or components can be bought that would allow students to take those kits home and work on them remotely – take into account whether every student needs each part or whether the tasks can, e.g., be split between hardware, software, etc., work.

Such equipment kits may be useful for in-person labs, too, and would allow maximum flexibility. Where at-home equipment is not possible, can existing benchtop equipment be configured via a remote link? Can videos of a TA/instructor doing the lab (and then uploading the data acquired) be substituted for some labs? Can software substitute for some hands-on labs?

TAs should run a planned and organized lab online just as they would in person. Consider if attendance should be mandatory or not. It may make sense to require students to check off their labs but may not make sense to require everyone to be present for the full lab session. Consider having in-person office hours (TA or instructor) to help students who are having problems. Create training and use videos for the labs that can be uploaded to Canvas. This may take more than one semester to create, so prioritize which are essential to get started and which may wait as time allows.

REFERENCES

1. L. D. Fink, *Creating significant learning experiences: An integrated approach to designing college courses.* John Wiley & Sons, Hoboken, NJ, USA, 2013.
2. S. A. Ambrose, M. W. Bridges, M. DiPietro, M. C. Lovett, and M. K. Norman, *How learning works: Seven research-based principles for smart teaching.* John Wiley & Sons, Hoboken, NJ, USA, 2013 2010.
3. R. M. Felder, "The Felder-Silverman model of Learning Styles." https://www.engr.ncsu.edu/stem-resources/legacy-site/learning-styles/. (accessed Aug. 18, 2019).
4. C.-M. Chan, A. Shamsuddin, and A. Suratkon, "Of Grades, Activities and Learning Styles: Correlation in a Civil Engineering Technology Course," *Advanced Science Letters*, vol. 24, no. 6, pp. 4576–4580, 2018.
5. R. M. Felder and J. Spurlin, "Applications, reliability and validity of the index of learning styles," *International Journal of Engineering Education*, vol. 21, no. 1, pp. 103–112, 2005.
6. M. Barak, R. Hussein-Farraj, and Y. J. Dori, "On-campus or online: examining self-regulation and cognitive transfer skills in different learning settings," *International Journal of Educational Technology in Higher Education*, vol. 13, no. 1, pp. 1–18, 2016.
7. T. L. S. Coulter, "*Improving Engineering Technology Curriculum Through the Identification of Effective Motivational Strategies and Teaching Approaches,*" presented at the *ASEE St. Lawrence Section Conference*, Cornell University, New York, Apr. 2018.
8. Y. Shi, H. Yang, J. MacLeod, J. Zhang, and H. H. Yang, "College Students' Cognitive Learning Outcomes in Technology-Enabled Active Learning Environments: A Meta-Analysis of the Empirical Literature," *Journal of Educational Computing Research*, vol. 58, no. 4, pp. 791–817, Jul. 2020, doi: 10.1177/0735633119881477.
9. E. J. Theobald et al., "Active learning narrows achievement gaps for underrepresented students in undergraduate science, technology, engineering, and math," *PNAS*, vol. 117, no. 12, pp. 6476–6483, Mar. 2020, doi: 10.1073/pnas.1916903117.
10. A. Misseyanni, P. Papadopoulou, C. Marouli, and M. D. Lytras, "Active learning stories in higher education: Lessons learned and good practices in STEM education," in *Active Learning Strategies in Higher Education*, Emerald Publishing Limited, Bingley, UK 2018.
11. "Is Passive Learning Unethical? The Science of Teaching Science | by It's About Time | Medium." https://medium.com/@ItsAboutTimeEDU/is-passive-learning-unethical-the-science-of-teaching-science-d5eb526506db (accessed Jul. 16, 2020).

12. C. H. Crouch and E. Mazur, "Peer Instruction: Ten years of experience and results," *American Journal of Physics*, vol. 69, no. 9, pp. 970–977, Sep. 2001, doi: 10.1119/1.1374249.

13. E. Mazur, *Peer Instruction: A User's Manual*. Pearson, New York City, New York, USA, 1996.

14. A. P. Fagen, C. H. Crouch, and E. Mazur, "Peer instruction: Results from a range of classrooms," *The Physics Teacher*, vol. 40, no. 4, pp. 206–209, 2002.

15. R. M. Felder and R. Brent, "Designing and teaching courses to satisfy the ABET engineering criteria," *Journal of Engineering Education*, vol. 92, no. 1, pp. 7–25, 2003.

16. K. Arshad and M. A. Imran, "Increasing the interaction time in a lecture by integrating flipped classroom and just-in-time teaching concepts," *Compass: Journal of Learning and Teaching*, vol. 4, no. 7, 2013.

17. S. E. Hizer, P. W. Schultz, and R. Bray, "Supplemental Instruction Online: As Effective as the Traditional Face-to-Face Model?," *J Sci Educ Technol*, vol. 26, no. 1, pp. 100–115, Feb. 2017, doi: 10.1007/s10956-016-9655-z.

18. C. M. Furse and D. H. Ziegenfuss, *"Teaching Flipped,"* 2016. http://www.teach-flip.utah.edu (accessed Dec. 05, 2016).

19. C. M. Furse and D. H. Ziegenfuss, "Co-flipped teaching: Experiences sharing the flipped class," in *2015 IEEE International Symposium on Antennas and Propagation & USNC/URSI National Radio Science Meeting*, 2015, pp. 1027–1028.

20. C. Furse, "A busy professor's guide to sanely flipping your classroom," in *Antennas and Propagation Society International Symposium (APSURSI), 2013 IEEE*, 2013, pp. 2171–2172.

21. D. Ziegenfuss and C. Furse, "Flipping the Feedback: The Value of Formative Assessment and Student Reflection in an Active Learning Flipped Freshman Engineering Class," *submitted to IEEE Transactions on Education*, 2019.

22. D. Ziegenfuss, C. Furse, E. Sykes, and E. Buendía, "Beyond the Click: Rethinking Assessment of an Adult Professional Development MOOC.," *International Journal of Teaching and Learning in Higher Education*, vol. 31, no. 1, pp. 63–72, 2019.

23. D. F. Ziegenfuss Cynthia, *"Evidence-Based Practice: Student-Centered and Teacher-Friendly Formative Assessment in Engineering,"* presented at the submitted to 2018 Annual American Society for Engineering Education Conference and Exposition, 2018.

24. C. M. Furse, D. Ziegenfuss, and S. Bamberg, "Learning to teach in the flipped classroom," 2014, pp. 910–911.

25. J. Bergmann and A. Sams, "How the flipped classroom is radically transforming learning," *The Daily Riff*, vol. 4, pp. 1–3, 2012.

26. J. L. Bishop and M. A. Verleger, *"The flipped classroom: A survey of the research,"* in *ASEE national conference proceedings*, Atlanta, GA, 2013, vol. 30, pp. 1–18.

27. R. Toto and H. Nguyen, *"Flipping the work design in an industrial engineering course,"* presented at the 39th *IEEE Frontiers in Education Conference*.

28. B. Osueke, B. Mekonnen, and J. D. Stanton, "How Undergraduate Science Students Use Learning Objectives to Study," *J Microbiol Biol Educ*, vol. 19, no. 2, Jun. 2018, doi: 10.1128/jmbe.v19i2.1510.

29. B. J. Beatty, *Hybrid-Flexible Course Design*. EdTech Books, edtechbooks.org, 2019.

30. C. M. Furse, *"ECE 3300 Introduction to Electromagnetics,"* 2016. https://utah.instructure.com/courses/224865 (accessed Dec. 05, 2016).

31. C. M. Furse, *"ECE 5340/6340 Numerical Electromagnetics,"* 2016. http://www.eng.utah.edu/~cfurse/ece6340/ (accessed Dec. 05, 2016).

32. C. M. Furse and D. H. Ziegenfuss, "A Busy Professor's Guide to Sanely Flipping Your Classroom: Bringing active learning to your teaching practice," *IEEE Antennas and Propagation Magazine*, vol. 62, no. 2, pp. 31–42, 2020.

33. J. Biggs, "Enhancing teaching through constructive alignment," *Higher education*, vol. 32, no. 3, pp. 347–364, 1996.
34. A. Dávila, Wiggins, G., & McTighe, J. (2005) *Understanding by design*. Alexandria, VA: Association for Supervision and Curriculum Development ASCD. Facultad de Ciencias y Educación de la Universidad Distrital, Bogotá Colombia, 2017.
35. R. B. Barr and J. Tagg, "From teaching to learning—A new paradigm for undergraduate education," *Change: The magazine of higher learning*, vol. 27, no. 6, pp. 12–26, 1995.
36. M. Weimer, "Learner-centered teaching and transformative learning," *The Handbook of Transformative Learning: Theory, Research, and Practice*, pp. 439–454, 2012.
37. M. F. Iskander, "Computer-based electromagnetic education," *IEEE Transactions on Microwave Theory and Techniques*, vol. 41, no. 6, pp. 920–931, 1993.
38. F. T. Arslan, *"Improved Flipped Classroom Teaching for Electromagnetic Engineering Course,"* presented at the *Proceedings of the 2015 ASEE Gulf, Southwest Annual Conference*.
39. M. B. Cohen and A. Zajic, *"Revitalizing electromagnetics education with the flipped classroom,"* presented at the *2015 USNC-URSI Radio Science Meeting (Joint with AP-S Symposium). IEEE*.
40. M. Stickel, "Teaching electromagnetism with the inverted classroom approach: Student perceptions and lessons learned," *AGE*, vol. 24, p. 1, 2014.
41. M. B. Cohen and A. Zajic, *"The Flipped Classroom Approach to Engineering Electromagnetics: A Case Study,"* presented at the *IEEE International Symposium on Antennas and Propagation and USNC-URSI Radio Science Meeting*, Atlanta, GA, Jul. 2019.
42. S. R. H. Hoole, S. Sivasuthan, V. U. Karthik, and P. R. P. Hoole, "Flip-teaching engineering optimization, electromagnetic product design, and nondestructive evaluation in a semester's course," *Computer Applications in Engineering Education*, vol. 23, no. 3, pp. 374–382, 2015.
43. "Our History | MIT OpenCourseWare | Free Online Course Materials." https://ocw.mit.edu/about/our-history/ (accessed Apr. 25, 2019).
44. "About Khan Academy | Khan Academy." https://www.khanacademy.org/about (accessed Apr. 25, 2019).
45. "Learn the Latest Tech Skills; Advance Your Career | Udacity." https://www.udacity.com (accessed Apr. 25, 2019).
46. "Coursera | Online Courses & Credentials by Top Educators. Join for Free," *Coursera*. https://www.coursera.org/ (accessed Apr. 25, 2019).
47. "edX," *edX*. https://www.edx.org/ (accessed Apr. 25, 2019).
48. "Canvas Network | Free online courses | MOOCs." https://www.canvas.net/ (accessed Apr. 25, 2019).
49. P. A. Cohen, B. J. Ebeling, and J. A. Kulik, "A meta-analysis of outcome studies of visual-based instruction," *ECTJ*, vol. 29, no. 1, pp. 26–36, 1981.
50. M. Prensky, "Digital natives, digital immigrants part 1," *On the Horizon*, vol. 9, no. 5, pp. 1–6, 2001.
51. D. Zhang, L. Zhou, R. O. Briggs, and J. F. Nunamaker Jr, "Instructional video in e-learning: Assessing the impact of interactive video on learning effectiveness," *Information & Management*, vol. 43, no. 1, pp. 15–27, 2006.
52. B. J. McNeil and K. R. Nelson, "Meta-analysis of interactive video instruction: A 10 year review of achievement effects.," *Journal of Computer-Based Instruction*, 1991.
53. "Product | IPEVO," *IPEVO (United States)*. https://www.ipevo.com/products (accessed Aug. 18, 2019).

54. Nalin, "Use Your Phone as a Streaming Document Camera," *The Very Spring and Root*, Aug. 15, 2013. http://www.theveryspringandroot.com/blog/2013/08/use-your-phone-as-a-document-camera/ (accessed Aug. 18, 2019).

55. C. Furse, "Lecture-free engineering education," *IEEE Antennas and Propagation Magazine*, vol. 53, no. 5, pp. 176–179, 2011.

56. R. M. Felder, R. Brent, and M. J. Prince, "Engineering instructional development: Programs, best practices, and recommendations," *Journal of Engineering Education*, vol. 100, no. 1, pp. 89–122, 2011.

57. M. Silverman, *Active Learning: 101 Strategies To Teach Any Subject*. ERIC, 1996.

58. R. M. Felder, "Richard Felder | Education-Related Papers | College of Engineering | NC State University." https://www.engr.ncsu.edu/stem-resources/legacy-site/education-related-papers/ (accessed Aug. 18, 2019).

59. R. M. Felder and R. Brent, *Teaching and learning STEM: A practical guide*. John Wiley & Sons, Hoboken, New Jersey, USA, 2016.

60. J. L. Jensen, T. A. Kummer, and M. Godoy, "Improvements from a flipped classroom may simply be the fruits of active learning," *CBE—Life Sciences Education*, vol. 14, no. 1, p. ar5, 2015.

61. M. Prince, "Does active learning work? A review of the research," *Journal of engineering education*, vol. 93, no. 3, pp. 223–231, 2004.

62. S. Freeman et al., "Active learning increases student performance in science, engineering, and mathematics," *Proceedings of the National Academy of Sciences*, vol. 111, no. 23, pp. 8410–8415, 2014.

63. R. M. Felder, D. R. Woods, J. E. Stice, and A. Rugarcia, "The future of engineering education II. Teaching methods that work," *Chemical Engineering Education*, vol. 34, no. 1, pp. 26–39, 2000.

64. A. Lakhtakia and C. Furse, *The World of Applied Electromagnetics*. Springer-Verlag, New York city, New York, USA, 2018.

65. T. Angelo and P. Cross, *Classroom assessment techniques*. San Francisco, CA: Josey-Bass, 1993.

66. C. Furse, N. Cotter, and A. Rasmussen, *"Bottlenecks and muddiest points in a freshman circuits course,"* 2018 Annual Amercan Society for Engineering Education Conference and Exposition, Salt Lake City, Utah, June 24–27, 2018.

67. P. Black and D. William, "Assessment and classroom learning," *Assessment in Education: Principles, Policy & Practice*, vol. 5, no. 1, pp. 7–74, 1998.

68. A. Shekar, "Active learning and reflection in product development engineering education," *European Journal of Engineering Education*, vol. 32, no. 2, pp. 125–133, 2007.

69. M. Rodgers, M. P. Grays, K. H. Fulcher, and D. P. Jurich, "Improving academic program assessment: A mixed methods study," *Innovative Higher Education*, vol. 38, no. 5, pp. 383–395, 2013.

70. J. J. Shultz and A. Cook-Sather, *In our own words: Students' perspectives on school*. Rowman & Littlefield, Lanham, Maryland, USA, 2001.

71. T. H. Macan, C. Shahani, R. L. Dipboye, and A. P. Phillips, "College students' time management: Correlations with academic performance and stress.," *Journal of Educational Psychology*, vol. 82, no. 4, p. 760, 1990.

12 Learning and Teaching in a Time of Pandemic

Hugo G. Espinosa
Griffith University, Australia

Uday Khankhoje
Indian Institute of Technology Madras, India

Cynthia Furse
University of Utah, USA

Levent Sevgi
Istanbul OKAN University, Turkey

Berardi Sensale Rodriguez
University of Utah, USA

CONTENTS

12.1   INTRODUCTION

The world has been severely impacted by COVID-19 in different social and eco-
nomic sectors, and the higher education sector is no exception. In the early stages of
the pandemic, several universities around the world closed their services temporarily
to help reduce the spread of COVID-19, while others immediately transferred the
delivery of all learning activities to online. Even universities with substantial experience in providing online learning had to adapt to the drastic challenges produced by
the pandemic [1].

Videoconference tools such as Zoom [2], Google Meet [3], and Microsoft Teams [4]
became the day-to-day resource for online teaching, and existing teaching tools such as
Collaborate Ultra [5] from Blackboard were employed to their maximum capabilities
to satisfy the online demand. No matter what resource was being used, online teaching
and learning became critical for the higher education sector, and it seems likely that
remote learning will not disappear post-pandemic. On the contrary, universities appear
to be modifying existing systems and processes in order to more permanently deliver
teaching activities online, as well as to widen the reach of their content.

Teaching in its traditional face-to-face form has attempted to move into more
innovative pedagogical strategies in past years, aiming to increase student
engagement, learning outcomes, and retention. With an emphasis on electromagnetics
engineering education, such strategies include flipped-classroom [6], experiential
learning [7], and project-based learning [8]. The challenge now is to learn how to
adopt and adapt these pedagogical strategies for use in an online teaching environment.
With the ease of COVID-19 restrictions in several countries, universities are adopting
hybrid learning strategies that allow for flexible learning and teaching models by
combining asynchronous and synchronous delivery, where students can undertake a
course online or in person [9]. Remote integrated platforms, laboratory simulation,
video experiment recordings, and mobile learning are some of the resources employed
by academics in a virtual learning environment [10]. One of the advantages of online
education is the possibility to collect digital metrics to improve student engagement,
teaching efficacy, and learning outcomes in online modes [11].

In this chapter, five academics representing four different universities and
geographical regions share their experiences of, and reflections on, teaching
electromagnetics during times of intense disruption, as was evident during the
COVID-19 pandemic. In most cases, the authors were required to transition their
traditional face-to-face classes to completely online or hybrid delivery modes. Their
reflections, as well as student feedback, on how they coped with the sudden transition
to online learning amid the additional pressures presented by COVID-19, are
discussed. The purpose of the chapter is to share the lessons learnt from this
experience and discuss implications for the future of engineering education. In
Section 12.2, each author shares their own specific reflections; although specific to

one cohort of students within a particular university, the lessons are easily transferable to other geographical regions and student groups. In Section 12.3, some final thoughts on the experiences are shared.

12.2 EXPERIENCES AND REFLECTIONS

12.2.1 THE DIGITAL DIVIDE AND POST-COVID-19 (CYNTHIA FURSE)

With great opportunities for remote education also come great challenges. One of the most serious of these is the Digital Divide. In many areas of the world, families (or even schools) have little or no Internet access and, perhaps, little or no access to cellular phone signals. As the shift to online learning during the 2020 pandemic showed, even in developed nations, significant portions of the population are without computers and Internet at home. This may be due to lack of infrastructure availability and limited connectivity for the infrastructure that exists (particularly in rural areas) or due to affordability when the infrastructure is available.

Students with limited or no Internet at home experience the following challenges:

- Most online tools such as learning management systems (Canvas, Blackboard, Moodle, etc.), Google and its apps, Office365, YouTube, Microsoft Teams, Zoom, e-books, and digital libraries, Wikipedia, and so many more all require Internet access.
- Completing online assignments and exams can be difficult or impossible. Imagine starting an online quiz with a 20-minute window with an unstable Internet connection.
- Synchronous learning activities such as group work, being active in classroom discussions, asking/answering questions may be impossible. Even asynchronous versions may be difficult or impossible. Even when recordings are available, it may be difficult to download or watch them.

Alternative solutions to overcome the previous challenges:

- Discuss with your students what may work best for them and do your best to accommodate student needs.
- Record everything and use automatic transcription options (such as in Zoom) when possible. A written record may be easier to download than a video. Enable students to do everything asynchronously, if needed.
- Be as flexible as possible with assignment due dates and be prepared to adjust if needed.
- Consider options that do not require stable and regular Internet connectivity. Create packets for download (or even put them on a DVD/USB and mail them) and make ways for students to upload assignments or email or mail them to you.
- Help students access Internet connectivity if possible. Many community service providers have special considerations for students. Libraries may check out laptops or other equipment.

What Students Want After COVID-19: In early 2020, education the world over was bumped from an in-person experience to online in a matter of weeks. While the unprecedented disruption was chaotic, it also presented an opportunity for faculty and students alike to gain new educational tools and skills. In mid-2020, we reviewed written student course feedback from the period of the transition and student comments from a university-wide survey, and interviewed faculty who had transitioned their teaching. The goal was to identify what students want -- key messages from students to help faculty better prepare for an online fall semester. We identified three key messages for a successful online/hybrid experience:

1. Clear Communication about expectations, assignments, learning objectives, and so forth;
2. Coordination between course elements, such as homework, laboratories, and exams all supporting the same concepts; and
3. Caring, that is, recognizing the learning and life challenges of the students and helping them work through them.

Perhaps not surprisingly, these messages are not unique to COVID-19 induced online classes. Rather, the abrupt transition to online amplified these challenges and raised the stakes for high-quality communication, coordination, and caring [12].

12.2.2 ONLINE TEACHING OF A LABORATORY-BASED COURSE (BERARDI SENSALE-RODRIGUEZ)

In response to the COVID-19 pandemic, the University of Utah decided to switch all its classes to a virtual format. This decision was made during the spring break (i.e., March 2020), thus halfway through the academic semester. Because of the urgency in which this decision was made, this change in instruction delivery format was announced on very short notice.

During this semester, we were delivering our undergraduate-level electromagnetics class: *Fundamentals of Electromagnetics and Transmission Lines*. This is a class with a strong laboratory component in which students put in practice the concepts discussed during the theory lectures. The laboratory is a semester-long design project that focuses on understanding the principles and developing the elements of a cardiac communications system [13]. The course laboratory gives students hands-on experience and puts the importance of what is learned in class in a real-world context. In this laboratory, students learn through measuring and modeling: the practical aspects and implementation of transmission lines, the dielectric properties of materials, and how these affect electromagnetic fields, antennas, and matching networks for these antennas, as well as link budgets in communication systems. This is a very well-structured laboratory experience that gives students a global perspective of electromagnetics, transmission lines, and its real-world applications.

When the university announced its intention to switch instruction modes, our biggest concern at the time, since the course has a strong laboratory component, was what to do with regard to the laboratory? To not significantly alter the student experience, we decided to provide each student with an at-home laboratory kit. The class was composed of approximately 20 students; thus, from an economical and practical perspective, it was possible to purchase, put together, as well as ship and deliver a laboratory kit to each student in a very short timeframe (two weeks). The current availability of low-cost test equipment, like the *nano-VNA*, can facilitate the implementation of at-home laboratories for electromagnetics and microwave courses [14]. Simple, low-frequency microwave circuits, such as matching networks, filters, and so forth, can be designed and fabricated by students at home using RF substrates and copper tape; this enables students to not just design and measure but also to fabricate these at home. Therefore, it is possible to provide students with an at-home laboratory kit for less than $200 (US dollars).

The laboratory kits we deployed consisted of the following items:

- SMA Male to Male 5ft cables
- Single-Sided Copper Clad Boards (FR4)
- Roll of Magnet Wire
- Roll of Foil Tape
- Pack of Standing SMA connectors
- Pack of Angle SMA connectors
- Soldering Iron Kit
- Rotary Tool
- USB Vector Network Analyzer (nano-VNA)
- UHF ham radio antenna (400-470 MHz)

We tried to remain as close as possible to each laboratory session's original objectives and keep things cost-effective, which was a challenge. It is to be noticed that the rotary tool and soldering iron could be eliminated if boards were prepared beforehand by lab staff and shipped with soldered connectors and drilled holes. However, because of the lack of time and scarcity of human resources, given the sudden transition to online instruction, we decided to provide all the elements to each student for putting together each experiment from scratch. Employing these kits, students performed three laboratories at home.

12.2.2.1 Laboratory Session 1: Microstrip Lines, Dielectric Materials, and Attenuation

The fundamental purpose of this laboratory is to become familiar with the operation of a vector network analyzer and employ this to characterize the properties of microstrip lines that students will design and fabricate on an FR4 dielectric substrate. A secondary goal is to learn about the dielectric properties of materials and how these properties affect electric fields. This is important for a cardiac communication system since communication needs to happen to/from inside the body. For this purpose, students use gelatin as an example dielectric that mimics the properties of the human body (skin, fat, muscle, when varying the water % and by adding salt into the mixture) and analyze the response of the microstrip lines fabricated under this new dielectric environment.

12.2.2.2 Laboratory Session 2: Monopole Antenna and Single Stub Matching Network

It is essential that the antenna for the pacemaker wireless communication system be matched to the transmission line, so that the majority of the signal is radiated rather than reflected by the antenna back to the source. Students learn about antennas, their input impedance, and how to match the antenna to a source using a single-stub matching network in this lab. Specifically, they create a matched equivalent half-wave dipole antenna to a 50-ohm microstrip line in an FR4 substrate.

12.2.2.3 Laboratory Session 3: Antenna Radiation Pattern, Effect of Dielectric Environment, and Communication Links

Students determine a link budget for a simplex (one-way) communication system for the cardiac pacemaker designed throughout the semester. The link budget predicts how much power is received from a known transmitter. The directivity and gain of the antenna, the propagation loss in the tissue simulation material, etc., are all to be considered. Students roughly determine the radiation pattern of the antenna they fabricated in the previous laboratory and explore the effect of the dielectric environment on the resonance properties of antennas.

Once we decided on the approach to be followed and the items that needed to be purchased, and had modified the original laboratory objectives, the next big challenge started: logistics. We did not just need to keep our costs within a reasonable budget, but also make sure that components would be delivered to us at the university as soon as possible, so we could assemble the individual laboratory kits and deliver them to each student.

We had received all items during the week after the spring break and started putting the laboratory kits together. We contacted the students so they could either collect their kits in-person from our stockroom or have their kits shipped directly to their homes. Around half of the students opted to pick them up, and the other half were either delivered by us to the students' homes or shipped through the mail. In two weeks, we had transitioned an in-person laboratory course that relies on sophisticated equipment and experiments into a fully at-home experience.

Once the laboratory kits were deployed, the next challenge arose: students having to perform the laboratories in their own homes. Despite this being the first time we had worked in a virtual capacity, the laboratories ran quite smoothly. One factor that contributed to this was the preparation undertaken before the laboratories. Each week, prior to the sessions, the course instructor would liaise with the teaching assistant (TA), who would first try the experiments at his home following the laboratory session guidelines step-by-step. Based on the TA comments and the issues the TA encountered, the instructor would modify the laboratory guidelines to address these challenges. This interaction was an important learning experience for both the TA and instructor, which enabled us to improve the laboratory guidelines and be highly efficient at the moment of interacting with students during the live laboratories. While it was possible for students to asynchronously complete the laboratories, there were standard deadlines for submitting laboratory reports. Approximately once or twice per week, the TA would hold virtual office hours through Zoom, where they would help students with the issues they experienced in the lab. Students were also able to contact the instructor through email, or during online office hours, with their questions about the lab. We found this asynchronous delivery of the laboratory worked well for the students.

Interestingly, a majority of students actually seemed to prefer the flexibility of this transformed delivery mode over the traditional method of attending traditional face-to-face, scheduled laboratories. This new at-home method was expanded to include all of the labs for this class (the ones mentioned here, as well as the ones previously taught prior to the COVID-19-induced shutdown) in fall semester 2020, as well as for all of the labs in our Microwave Engineering I class. Similarly, at-home labs were developed for our freshman and sophomore circuits classes. At the University of Utah, a significant fraction of undergraduate students have regular full-time jobs. Permanently having the option of students being able to do the laboratories at home and work these asynchronously provides a flexibility that is important for such students.

In conclusion, our experience shows that it is possible to deliver an electromagnetics lab-based course remotely by providing at-home laboratory kits to students. The current availability of low-cost equipment can transform RF and electromagnetics education due to the possibility of realistically performing experiments at home that until recently would have required a dedicated laboratory setting. The cost of such equipment is continually decreasing, and functionality and performance are dramatically increasing. Furthermore, the availability of open-source hardware and software enables a huge degree of flexibility. Although our experience in deploying an at-home electromagnetics laboratory course was on a relatively small/middle-sized class, this approach is scalable for smaller and larger class sizes.

12.2.3 EXPERIENCE OF FACE-TO-FACE AND ONLINE TEACHING (LEVENT SEVGI)

Our 2019–2020 spring semester commenced in early February 2020; after just five weeks of face-to-face lectures, teaching activities were moved online in response to the COVID-19 pandemic. The potential adverse impact of this disruption was mitigated, however, due to my university (Istanbul Technical University) already possessing an effective technological infrastructure to support online teaching activities and a well-organized Distance Learning Center. Similarly, many universities are increasingly supplementing traditional face-to-face learning models with online, hybrid, and blended learning components. This allows students to participate in high-quality learning situations from anywhere at any time. It also enables more dynamic interactions between the lecturer and students and among the students themselves. Having access to the appropriate infrastructure, resources, and experience in online teaching, undoubtedly assisted many educators with the sudden transition to fully online teaching resulting from COVID-19.

Nonetheless, all educators naturally approached the sudden and unprecedented shift to fully online teaching with varying degrees of enthusiasm and concern. Some were very optimistic in light of the advantages of online learning (e.g., flexibility, affordability, efficiency, ease of access, increased quality of learning material, etc.);

others were more skeptical about whether these advantages outweighed the potential disadvantages of online learning (e.g., motivation loss, self-discipline problems, less social interaction, etc.). Furthermore, the sudden transition from face-to-face to online teaching presented several challenges for both educators and students. For instance, the success of an online program initially depends on technology and computer literacy of both the educators and the students: they must have access to the online learning environment and must possess a minimum level of computer and Internet knowledge in order to function successfully in that environment. Much of the success also depends on lecturer creativity, and the students' level of ability, experience, and willingness to learn. The immediacy of the shift to online teaching exacerbated these challenges for some people.

My first class to transition fully online this year was an undergraduate electromagnetics course - *Electromagnetics Wave Theory* - with nearly 30 students enrolled. I felt well-prepared for online teaching due to my years of experience in teaching electromagnetics courses, which include electromagnetics wave theory, antennas and propagation, electromagnetics scattering and diffraction, and introduction to electromagnetic compatibility (EMC) [15–22]. Furthermore, I had also developed digitized teaching materials (e.g., lecture notes, quizzes, exams, and presentations) and online teaching modules (e.g., video recordings of lectures). Nevertheless, I encountered several minor challenges in the first few online sessions for this course, including low-quality, low-speed Internet connections, disruptions in audio and video playback, miscommunications with students, and issues with the first few assessment pieces. After a few trials, I was able to resolve these issues, resulting in useful and productive experiences with the online lectures and assessments.

The novel and unprecedented situation experienced by educators in response to the COVID-19 pandemic has renewed interest in developing new, or adapting existing, online teaching methods to support students' learning, engagement, and retention. In relation to electromagnetics, for example, the use of virtual learning tools addresses many of the challenges faced by educators within this field, including conveying complex theory, providing hands-on experience to communicate the practical application of theory, and adapt teaching methods to meet students' needs.

Currently, electrical engineering education requires a range of multidisciplinary, physics-based, problem-matched analytical and computational skills. In electrical engineering, real-life systems (from nanoscale to km-wide) are among the most complex ones and are a fundamental part of electromagnetics. We, therefore, have witnessed the transformation from engineering electromagnetics to electromagnetic engineering for the last several decades.

Electromagnetic theory is well-established with Maxwell equations, but teaching/ lecturing is always a challenge. Experimentation, theoretical background, and numerical simulations are critical. An intelligent approach is to simultaneously use physics-based modeling, hands-on training, numerical-based modeling, and computer simulations. Electromagnetic problems may generally be grouped into three categories: (a) Antenna and Radiation problems, (b) Scattering and Diffraction Problems, (c) Waveguiding structures. In order to deal with all these problems, one needs to start with Maxwell equations, either in differential or in integral form, and state the boundary conditions (i.e., introduce the geometry).

Furthermore, it is important for educators to adapt their teaching to meet the needs of students with varying levels of knowledge, experience, skills, and motivation. It may be speculated that more capable and motivated students require minimal guidance to succeed, regardless of whether they are studying face-to-face or online. For these students, clearly stating and defining real-life problems in lectures and summarizing how these problems are mathematically modeled may provide them with sufficient direction to understand the problems, learn how to use mathematical relations, and find the solutions. In contrast, for students who have difficulties understanding real-life problems and applying mathematical relations (e.g., differential equations, double and triple integrals, infinite and poorly convergent series summations, systems of equations, nonlinear equations, etc.) starting with Maxwell equations and boundary conditions may really be a challenge. What we have done for these students is to use numerical modeling and simulation plus visualization. This was our initial motivation for developing and using electromagnetic virtual tools in teaching/training next generations, especially for those with less knowledge and skills [15–17].

We have developed and introduced multipurpose electromagnetics virtual tools that can be used in most of the classical electromagnetics lectures as well as in novel Electromagnetics Engineering Programs [18–22]. Students first interact with these virtual tools and visualize time and frequency domain results for all different types of electromagnetic problems. They learn the value of Maxwell equations and the physical explanations behind these mathematical equations. This enables them to understand the relation between physics and mathematics and have more confidence in engaging with the course content.

Virtual tools are increasingly being integrated in engineering education to support the learning of future generations of electromagnetics researchers and students (see Chapter 5). The overall effectiveness of such tools is very positive, and these tools continue to be employed and adapted. The impact on students is that they are better equipped to understand and apply mathematical expressions. When using virtual tools, it is important to consider their ease-of-use, whether the user has the ability to change some parameters while others remain fixed (to prevent the student from becoming lost), and finally, whether they are open to nonphysical results to some extent so that the user may comment on them. Some comments on novel engineering education curricula may be listed as:

- Novel education techniques such as problem-based learning, inquiry-based teaching should be discussed in detail under available teaching resources (i.e., personnel, laboratories, classroom capacities, number of students, etc.), and smooth adaptation should be made accordingly.
- Subjects like energy sources and conversion, environment, electromagnetic radiation, light and sound, thermodynamics, material science should be included in engineering curricula.
- Engineers should be well equipped to deal with data acquisition, correlation, models, epidemiology, risk management, information-based decision-making, uncertainty, and bounds of science.

- Basic lectures, such as measurement techniques, should be improved accordingly and concepts like accuracy, precision, resolution should be well taught. Error analysis is also another important concept that should be well taught.
- Hands-on practice and training are a must in engineering education. Although laboratories and test instruments have been simulated as virtual reality environments, which may be as good as the real environment, students still need hands-on training.
- Computer simulations are as necessary as hands-on training; therefore, modeling and simulation lectures should be included in engineering programs.
- Engineers become more publicized in parallel to technological developments. Therefore, lectures like Science Technology and Society, or Public Understanding of Science should be included in the engineering curricula

In conclusion, virtual tools assist in overcoming the challenges of teaching electromagnetics courses, in both face-to-face and online settings. The use of virtual tools, together with theoretical lectures, assists in helping students to understand and apply complex mathematical and scientific theory. This is especially important for student groups with varying levels of capability and motivation to learn.

12.2.4   USING A FLIPPED CLASSROOM IN AN ALL ONLINE MODE (UDAY KHANKHOJE)

During the July-November 2020 semester at the Indian Institute of Technology Madras, the undergraduate course on *Engineering Electromagnetics* was taught as a "flipped" classroom in an all-online mode, courtesy of the disruption introduced by COVID-19. The class consisted of approximately 140 sophomore students and was conducted by two instructors, Deepa Venkitesh and Uday Khankhoje, and ably assisted by 12 teaching assistants (TAs) of the Department of Electrical Engineering. The content was spread over 12 weeks, with a weeklong break mid-way.

Following a rigorous weekly schedule for the course, activities was found to be the key ingredient for the successful conduct of this course. The structure for each week of content included:

- Day 1: Release of lecture content, and problem set for the week.
- Day 6: Conducting problem-solving sessions with TAs, split across six online classrooms each having 20-25 students and two TAs. During these sessions, the final answers were not given; instead, the TAs encouraged students to enumerate their attempts and were given hints toward the arriving at the correct solutions. The use of collaborative online "whiteboards," with the TAs using suitable writing equipment such as digital pens (i.e., Apple Pencil, Wacom Pen, or any stylus pen for touch screen devices), made the sessions very effective. These sessions were recorded, and the recordings were made available to all students.
- Day 7: A live interactive lecture held by the instructors to summarize the week's contents, address students' questions or concerns, and provide a brief preview of the upcoming week's content. A recording of the lecture was also

shared with the students. In addition, the solutions to the problem set were released on this day.

- Day 8: A (brief) auto-graded quiz on the week's content was conducted. Additionally, the students were able to complete a feedback form for the week, which recorded how much of the content they were able to watch or understand, and if they had any specific difficulties during the week. The solutions to the quiz were uploaded the same day.

Behind the scenes, the instructors and TAs would also have regular weekly meetings to discuss the problems to be given for that week. The use of collaborative web-based tools for editing documents made this task smooth, as compared to people emailing documents back and forth. Additionally, individual TAs would report on the progress of the students in their respective TA rooms (used on Day 6) during this meeting. Each TA also formed a WhatsApp group with approximately 12 students for the duration of the semester. This helped the students immensely in clearing what they imagined to be "silly" doubts within a quick turn-around-time.

A special note needs to be made about the content management system (CMS) used to execute the course (we used the open-source tool, Moodle). In addition to providing a forum where all course contents, relevant links, and so forth could be posted, we also created a course discussion forum where students could ask questions about the content, or about any other aspect of the course. We found that this provided a rich learning experience for the students. Perhaps the most crucial aspect of the CMS was the ability to create custom question banks for the execution of the weekly quiz. For multiple-choice questions, the CMS allowed each student to see a shuffled version of the options. Additionally, for questions with numerical values, the CMS generated these numbers at random (with instructor specified ranges and distributions). As a result, with high probability, no two students would get the same question set for the weekly quiz. Choosing only objective-type questions (including those with numerical answers) allowed for the quiz to be auto-graded -- a critical feature for courses with large enrollments. This feature also greatly helped the learning process of the students, since they received feedback regarding their performance immediately after the end of the quiz. The maturity of the CMS is also key and should have reasonably functional interfaces for operating via a mobile phone interface.

In subjective feedback collected from the students, it was learnt that the weekly quiz feature, though slightly stressful, helped them to keep sync with the course progression. In courses where such a feature was not present, often a backlog of lectures to be watched resulted. This led to them losing interest in the course all together; therefore, the need for a regular course assessment seems essential in "taking the class along." In all, there were 12 weeks of content, and thus 12 weekly quizzes. It was made known to the students at the start of the semester that not all quizzes would count toward their final grade and, for example, the best 8 out of 12 approach would be taken. This a priori knowledge also helped in keeping student stress at bay. For the more subjective and longer end-of-semester examination, it is planned to conduct a similar online based exam, additionally allowing students to upload their worked-out solutions in order to receive partial credit.

Having the lecture content available prior to the start of the course made the planning process much easier for the instructors. Since 2003, Indian academia has pioneered the production of high-quality technical content in the Sciences and Engineering under the banner of "National Programme on Technology Enhanced Learning," or NPTEL [23]. Semester-long lecture recordings for a wide range of undergraduate and graduate-level courses have been posted online for free. For our offering, we used the lectures recorded by IEEE Fellow Prof. R. K. Shevgaonkar on the subject of Electromagnetic Waves [24]. An undergraduate textbook written by the same lecturer served as the reference book for the course [25]. It was very helpful to have both the lecture content and the reference book produced by the same person, as the lectures served to enunciate a thought flow consistent with the reference book. Having the same notation and conventions was an added bonus. Finally, each lecture was accompanied by a set of approximately five, very simple, objective-style questions within the CMS to assist in the retention of the subject matter. These questions were not graded and, hence, students faced no pressure in attempting them.

In a subject like Electromagnetics, simulations and visualizations can play a big role in understanding the content. During many of the live online lectures, the instructors would share simple simulations written in MATLAB, along with a demonstration of their output. The code would also be shared with the students for offline "play." Additionally, a few standalone lectures on MATLAB were recorded and shared with students, for instance, on visualizing standing waves on a transmission line, or on how to perform stub matching on a transmission line by using root-finding functions, rather than Smith charts. Although the course was not intended to focus on computational aspects, introducing students to MATLAB simulations in parallel helped them to gain confidence in the subject matter. The accompanying website [26] of a classic undergraduate textbook on Electromagnetics by Ulaby et al. also contains simulations of various phenomena, such as wave polarization and antenna visualizations, for instance, which were also demonstrated to the students along the way.

Thus, apart from just the recorded lectures, a lot of rich material was made available to students throughout the semester. During the mid-semester break, colleagues who work in the general area were requested to give guest lectures on their research area and reflections on Electromagnetics in general. This helped students make the mental transition from textbook knowledge to that of the real world. All of these experiences coupled with regular problem solving with the TAs and also in a "test" environment made for a good learning experience for the students.

A majority of the implementation challenges faced were on account of poor Internet access faced by many students. The majority of Indian students have 3-4G connectivity but many places, for example, in remote villages, might not even have 2G Internet access. In some such cases, students made periodic visits to nearby places with better Internet connectivity in order to download the week's content. Thus, it was important to make sure that all the content we made available was easily downloadable and compact in size, be they lecture recordings or problem set PDFs. During the course of the semester, there were a few extreme weather events in certain parts of the country, such as very heavy rainfall and resultant floods and power outages. During such a time, the weekly quizzes were suspended and held in a later week.

12.2.5 TRANSITION FROM FACE-TO-FACE TO BLENDED LEARNING DUE TO COVID-19 (HUGO G. ESPINOSA)

The higher education sector has faced numerous challenges to their ability to ensure the continuity of learning in a context of unprecedented disruption, instigated by the global pandemic. Many universities around the world have delivered courses online for some time; for them, the transition to online delivery due to COVID-19 has been relatively smooth. For other universities that have relied on more traditional face-to-face teaching methods, the transition to online delivery has brought about significant challenges. At Griffith University, Queensland, Australia, the practice of blended learning has existed for several years. In 2015, the Griffith Sciences Blended Learning Model was created [27], aiming to provide a framework to support the development of blended learning initiatives through the use of technology, and a combination of different pedagogical strategies. Some courses have adopted different initiatives for a blended experience to enhance learning and teaching; these include, for example, flipped classrooms and the use of PebblePad ePortfolios[1] to support students in planning, reflecting, sharing, and providing feedback. The latter has been successfully applied to first-year common engineering courses such as *Engineering Design Practice*.

Some discipline-specific courses from second year and above have not been redesigned in line with the blended learning model, and still rely predominantly on traditional face-to-face teaching activities. This is the case with the second-year engineering course *Electromagnetic Fields and Propagating Systems* from the Bachelor in Electronic Engineering at Griffith University. The course is divided into four modules: Electrostatics, Magnetostatics, Dynamic Fields, and Transmission Lines. In addition to the lecture material, the course contains six hands-on practical laboratory sessions that aim to solidify the course content. The laboratories include topics on electric and magnetic field mapping, standing waves, traveling waves, and impedance matching.

During Trimester 2 (T2; July to October 2020), the course was delivered in a blended mode with 35 students enrolled. Lectures were all delivered online, while laboratories were delivered on-campus. By September 2020, Australia and, in particular, Queensland, had registered a very low number of COVID-19 actives cases, and permissions were granted by the School of Engineering at Griffith University to conduct laboratories, tutorials, and workshops on-campus, while lectures remained online only.

Traditionally, the course assessment includes two quizzes (10% each), six laboratories (30% in total), and a final written exam (50%), all of which are usually conducted on-campus. In response to the change to online teaching, necessitated by the COVID-19 pandemic, the course assessment plan was modified. While the laboratories were still assessed on-campus, the final exam was conducted online and individualized based on each student's ID number to reduce plagiarism and cheating. The quizzes were converted into an oral assessment and an assignment; this modification was necessary to comply with the university's policy for academic integrity that, at the time of the pandemic, mandated a maximum of 50% of a course's total assessment be conducted online.

The selection of the learning management system was important; although both Microsoft Teams and Collaborate Ultra were available, Collaborate Ultra was the platform ultimately chosen for the online lectures. Blackboard is the learning management system used at the university, and Collaborate Ultra is part of Blackboard; therefore, the use of Collaborate Ultra allowed students to deal with only one platform. All course materials and communications, such as lecture notes, announcements, laboratory worksheets, e-mails, online assessments, and assignment submission points were all contained within the Blackboard system for the course.

The lecture material was made available to students prior to the start of each module. In addition, a set of non-assessed tutorial problems, relevant to each module, was made also available at the start of the module. Throughout the lectures, selected problems from the tutorial set were solved in class using a tablet. At the end of each module, solutions to all tutorial problems were released to students. The course uses the book *Fundamentals of Applied Electromagnetics*, by Fawwaz T. Ulaby and Umberto Ravaioli [28], as the main textbook.

The delivery of online lectures consisted of conceptual theory, practice problems, and, in some circumstances, synchronous demonstrations to solidify the mathematical and physical concepts. These demonstrations allowed for more interaction and engagement with students. Taking advantage of the online delivery, it was relatively easy to conduct real-time demonstrations under the camera and discuss them with students. Some examples included demonstrations of

- Oersted's discovery to show the connection between electricity and magnetism by using a current-carrying wire and a compass.
- Faraday's Law to show electromagnetic induction by moving a permanent magnet into a coil of wire. The coil was connected to an analog ammeter to measure the changes in current flow.
- Lenz's Law to understand the direction of the current induced by a changing magnetic field. The demonstration consisted in moving a neodymium permanent magnet near an aluminum can at different positions. This experiment led to the explanation of how nonferrous materials are sorted in the recycling industry.

Due to the nature of the laboratories and the necessity for using specialized equipment, such as oscilloscopes, function generators, magnetic field sensors, variacs, and slab-lines, students were required to complete the practical experiments on-campus. All laboratories were fully attended.

At the end of the trimester, anonymous student feedback was collected via a qualitative and quantitative survey. The purpose of the survey was to assess students' satisfaction with the course, the perceived effectiveness of the blended delivery mode, and identify implementation challenges perceived by the students. Thirty-one students participated in the survey, which contained the questions listed in Appendix 12.1. For questions 1 to 5, students were given the option to select *not at all*, *slightly*, *moderately*, or *very*. Figure 12.1 summarizes the responses for those five questions.

Students were mostly satisfied with the blended delivery of the course; most of them rated online learning as being an effective teaching mode and the majority were

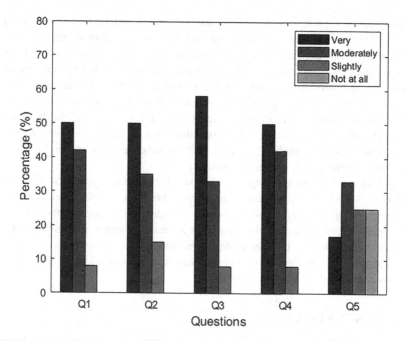

FIGURE 12.1   Anonymous survey responses from 31 students, questions 1 to 5.

satisfied with the technological resources available to them. Most students perceived that face-to-face communication was critically important during remote learning. Contrary to expectations, not all students were stressed about learning online and they seemed to have adapted to the format well.

In terms of the implementation resources, all students had access to a device for online learning for the duration of the course. They all used either a laptop or a desktop computer to follow the remote teaching. Around 60% of the students had excellent connectivity, while 40% had good connectivity, which means that Internet resources did not interfere with their learning experience. This metric, however, can vary depending on socio-demographics around the country.

Although 80–90% of the students reported that their online learning experience was effective and were very satisfied with online education in T2 2020, 70% would prefer both lectures and laboratories to be held on-campus in 2021. Only 30% would prefer to go back on-campus for the laboratories and keep the lectures online. While students seemed to have had a good online learning experience, they perceive more value in going back on-campus for a face-to-face delivery mode.

All students considered laboratories on-campus to be necessary for their learning; some even emphasized that "it is absolutely imperative that laboratories for this subject be held on-campus." It should be noted that the students included in the sample for this survey were enrolled on-campus and are used to on-campus delivery: if the laboratories were to be redesigned for online delivery, it is likely students would adapt to them well.

Regarding question 12, students reported that learning electromagnetics theory online was not overly difficult. Furthermore, they commended the way in which the content was presented, the interactivity through practical demonstrations, examples of real-life applications, lecture material, and explanations of theoretical principles. Some of them, however, recognized that isolation, uncertainty, and their family/home environment may have adversely affected their learning experience in some instances.

Finally, in terms of how different the online delivery of this course was with respect to other courses, students positively commented on the interactive delivery, course organization, use of computer resources, and, once again, the laboratories being held on-campus. However, they also added that the on-campus format allowed them to get more personalized support, learn from other students, and have more effective communication with lecturers and peers.

The way in which engineering education will further develop in a post-COVID-19 context is uncertain; what does seem clearer, however, is that blended learning methods seem to be an effective alternative to traditional classroom-based learning. The experience of delivering this course in a blended format, in conjunction with students' survey results, indicate that it is important to assess students' needs, from access to resources to their drive and motivation, when considering the design and delivery of blended learning models.

12.3 FINAL REMARKS

The disruption caused by COVID-19 forced many educators to move learning activities online, regardless of their level of experience with, or the suitability of their courses for, online, blended, or hybrid modes of delivery. This produced significant pedagogical and logistical challenges, for example, how to convey complex mathematical ideas, how to impart laboratory-based education, how to conduct course evaluations in a fair manner, and so forth. It has also brought into the spotlight the various problems faced by students learning online, for example, how to cope with having to spend most waking hours in front of an electronic screen, keep pace with a course if Internet access and power supply are not consistent, learn without having peer huddles, learn when the only electronic device they have is a smartphone, and balance their educational needs with family responsibilities. Finally, both educators and students alike experienced unexpected challenges relating to communicating across different time zones as a result of the geographical dispersion that occurred in response to COVID-19.

In the traditional face-to-face university setting, most of these problems did not exist and the focus was purely on learning. Thus, in the given setting, it would be a mistake for educators to go about "business as usual" and try to cram all the content that they would usually cover in a traditional teaching period, into an online offering. The content, and the form in which it is made available to students, needs to be carefully considered with respect to potential student constraints. For example, where possible, it is preferable to employ text or reference books that have electronic versions and software that can be run in a browser rather than software relying on large download/ heavy hardware requirements. In the case of students conducting laboratory experiments at home, it is imperative to reduce the cost of components or even consider the option of the university lending experimental kits to students, to be returned at the end of the semester.

In this chapter, we have seen various examples of the efforts made by educators across the globe to rapidly adapt their face-to-face teaching activities to an online environment in response to the disruption provoked by the COVID-19 pandemic. Some common themes emerge, for example, the use of virtual tools and simulations to convey the intuition behind the many concepts in electromagnetics, or the extensive use of content management systems and online tools to administer problem sets and quizzes. In those cases where it was possible to conduct laboratory experiments (either at home or at the university), it was found to greatly assist in the learning process.

APPENDIX 12.1

Survey questions for subchapter 12.2.5 (Figure 12.1) - Transition from face-to-face to blended learning due to COVID-19.

1. How would you rate your overall satisfaction with online education in Trimester 2 2020?
2. How effective has online learning been for you?
3. Are you satisfied with the technology and software (Blackboard, Teams, etc.) used for online learning?
4. How important is face-to-face communication for you while learning remotely?
5. How stressful was online learning in Trimester 2 2020?
6. Did you have access to a device for online learning in Trimester 2 2020?
   a. Yes and it worked well
   b. Yes but it did not work well
   c. No, I shared with others
   d. No, I used the computers at the university
7. What device did you use in Trimester 2 2020 for online learning? You can choose more than one answer.
   a. Laptop
   b. Desktop
   c. Tablet
   d. Smartphone
   e. Other (specify)
8. How would you rate your Internet connectivity during the online courses in Trimester 2 2020?
   a. Poor
   b. Average, I missed some lectures due to my connectivity
   c. Good, I missed part of a lecture due to my connectivity
   d. Excellent
9. Would you prefer going back on-campus in 2021?
   a. No, online delivery is fine
   b. Yes, but only for the laboratories, lectures online are fine
   c. Yes, I prefer both lectures and laboratories on-campus
   d. Other (Specify)
10. Did you enjoy having the laboratories for Electromagnetic Fields on-campus in Trimester 2 2020?

a. No, I would have preferred having the laboratories online
b. Yes, laboratories on-campus were necessary for my learning
c. Other (specify)
11. Please comment briefly on how easy it was for you to learn Electromagnetic Field theory online.
12. How different was the online delivery of this course (Electromagnetic Fields) with respect to your other online courses? Please briefly explain your response.

NOTE

1 PebblePad is a web-based platform that allows staff and students to create a personal learning environment (PLE). Students can plan their work, collect and create evidence of their learning, and reflect on their study, all in the form of eportfolios (https://www.pebblepad.com.au/).

REFERENCES

1. A. Schleicher, "The impact of COVID-19 on education: Insights from education at glance 2020," 2020. [Online]. Available: https://www.oecd.org/education/education-at-a-glance/. [accessed: Nov 15 2020].
2. A. Johnson, *Online teaching with Zoom*, Johnson, 2020.
3. Google.com, "Google for education products," 2020. [Online]. Available: https://edu.google.com/intl/en_au/teacher-center/products/?modal_active=none. [Accessed Nov 23 2020].
4. Microsoft.com, "Microsoft teams for education," 2020. [Online]. Available: https://www.microsoft.com/en-au/education/products/teams. [Accessed: Nov 15 2020].
5. Blackboard.com, "Blackboard collaborate with the ultra experience," 2020 [Online]. Available: https://help.blackboard.com/Learn/Instructor/Interact/Blackboard_Collaborate/Collaborate_Ultra. [Accessed Nov 15 2020].
6. C. M. Furse and D. H. Ziegenfuss, "A busy professor's guide to sanely flipping your classroom," *IEEE Antennas and Propagation Magazine*, vol. 62, no. 2, pp. 31–42, 2020.
7. H. G. Espinosa, T. Fickenscher, N. Littman and D. V. Thiel, *"Teaching wireless communications courses: An experiential learning approach,"* EuCAP 2020 (online), 14th European Conf. Antennas and Prop., Copenhagen, Denmark, March 2020.
8. A. Nieves, J. Urbina, T. Kane, S. Huang and D. Penaloza, *"Work in progress for developing project-based experiential learning of engineering electromagnetics,"* ASEE Annual Conference & Exposition, 2019.
9. B. Beatty, *"Transitioning to an online world: Using HyFlex courses to bridge the gap,"* EdMedia+ Innovate Learning. Association for the Advancement of Computing in Education (AACE), 2007.
10. F. Darby and J. M. Lang, *Small teaching online: Applying learning science in online classes*, Jossey-Bass; San Francisco, CA, USA: Wiley, 2019.
11. C. Dede, J. Richards and B. Saxberg, *Learning engineering for online education*, New York, USA: Routledge, 2019.
12. E. Pronovost, I. Fuller, S. Lakey, D. Ziegenfuss and C. Furse, "Enhancing Education After COVID," 2020. [Online]. Available: https://www.youtube.com/watch?v=osk3I4v-2G0&list=PLbixl6t6CLKydOV02R3XDmu7c1YiW5nT6&index=17. [Accessed Nov 15 2020].

13. C. Furse, L. Griffiths, B. Farhang, and G. Pasrija, "Integration of signals/systems and electromagnetics courses through the design of a communication system for a cardiac pacemaker," *IEEE Antennas and Propagation Magazine*, vol. 47, no. 2, 2005.
14. Nanovna.com, "Handheld vector network analyzer," 2020. [Online]. Available: https://nanovna.com/. [Accessed: Nov 23 2020].
15. L. B. Felsen, L. Sevgi, "Electromagnetic Engineering in the 21st Century: Challenges and Perspectives," *ELEKTRIK, Turkish J. of Electrical Engineering and Computer Sciences*, vol. 10, no. 2, pp. 131–145, 2002.
16. L. Sevgi, *Complex Electromagnetic Problems and Numerical Simulation Approaches*, IEEE Press – Wiley, Piscataway, NJ, 2003.
17. L. Sevgi, *Electromagnetic Modeling and Simulation*, IEEE Press – Wiley (EM Wave Series), NJ, April 2014.
18. L. Sevgi, "EMC and BEM Engineering Education: Physics based Modeling, Hands-on Training and Challenges," *IEEE Antennas and Propagation Magazine*, vol. 45, no. 2, pp. 114–119, 2003.
19. L. Sevgi, C. Goknar, "An Intelligent Balance between Numerical and Physical Experimentation," *IEEE Potentials Magazine*, vol. 23, no. 4, pp. 40–44, 2004.
20. L. Sevgi, "Virtual Tools/Labs in Electrical Engineering Education," *ELEKTRIK, Turkish J. of Electrical Engineering and Computer Sciences*, vol. 14, no. 1, pp. 113–127, 2006.
21. L. Sevgi, "Electromagnetic Modeling and Simulation: Challenges in Validation, Verification and Calibration," *IEEE Trans. on EMC*, vol. 56, no 4, pp. 750–758, 2014.
22. L. Sevgi, "Teaching Electromagnetic Modeling and Simulation as a Graduate-Level Course," *IEEE Antennas and Propagation Magazine*, vol. 54, no. 5, pp. 261–269, 2012.
23. Nptel.ac.in, "National Programme on Technology Enhanced Learning," 2020. [Online]. Available: https://nptel.ac.in/. [Accessed: Nov 15 2020].
24. Nptel.ac.in, "Introduction to EM waves and various techniques of communication," 2020. [Online]. Available: https://nptel.ac.in/courses/117/101/117101056/. [Accessed: Nov 15 2020].
25. R. K. Shevgaonkar, *Electromagnetic waves*," Tata McGraw-Hill Education, 2005.
26. F. T. Ulaby, E. Michielssen and U. Ravaioli, *Fundamentals of Applied Electromagnetics*, 2010. [Online]. Available: http://em.eecs.umich.edu/. [Accessed: Nov 15 2020].
27. C. N. Allan, C. Campbell and J. Crough, *Blended learning designs in STEM higher education*, Singapore: Springer, 2019.
28. F. T. Ulaby and U. Ravaioli, *Fundamentals of Applied Electromagnetics*, 7th ed. New York, Essex, England: Pearson, 2015.

13 Conclusion and Outlook

Karl F. Warnick

Brigham Young University, USA

Krishnasamy T. Selvan

Sri Sivasubramaniya Nadar College of Engineering, India

Since the early 1900s, societal pressures, technological advances, and new delivery methods have influenced change in engineering education [1]. This has affected all subdisciplines, including traditional electromagnetics curricula. To guide EM education in this time of transition, we have attempted to provide a vision for how trends in higher education will cause the tools, practices, and curricular structure used in EM education to evolve. As changes occur, rather than hold to specific practices, it can be valuable to focus on guiding principles. Moving forward, EM teaching could be assimilated into curricula oriented toward quantum applications and materials or morph in any number of ways, but the framework and ideas in this book will remain valuable in helping teachers and students in the field.

Guiding principles for EM education as it evolves include appreciation for quality teaching and prioritizing learning over teaching alone. The current generation of students "like independent, self-paced learning, with opportunities for collaboration as needed. They see their instructor as a learning facilitator, who helps them to develop relevant and practical skills" [1]. The educator's role in such a scenario involves (1) developing and practicing teaching styles in consonance with modern learning styles of students, (2) nurturing the subject such as to facilitate the development of future contributors to the field, and (3) promoting the subject and its practice by effective use of professional societies.

While these three aspects are related, this book has primarily focused on (1) and (2) above. We addressed the requirements of (1) by considering experiential learning in Chapters 3 through 6. Experiential learning has been shown to significantly help with enhancing cognitive skills of students [2]. To navigate the 21st century global societal needs, graduates need empathy, ethical considerations, creativity, and life-long learning ability [1,2]. These skills, associated with the idea of what has become a popular term in industry, "design thinking," can be facilitated by experiential learning opportunities [2]. In addition to experiential learning, students will continue to be edified by the rich conceptual and theoretical content of electromagnetics. Various conceptual elements and approaches were discussed in Chapters 7 to 9.

In respect of item (3), organizational units of professional societies, such as the IEEE Antennas and Propagation Society and its Chapters, can work toward enhancing the quality of teaching and learning of EM in their regions [3]. Another way could be taking the subject beyond the electrical engineering domain, as discussed in Chapter 10.

An important principle of successful teaching and learning that underlies all these activities is teacher enthusiasm. Teacher enthusiasm is central to providing an effective educational experience to students. While techniques can add value, it is enthusiasm that infuses life into our classes. The importance of this is something the authors continue to experience in their academic career. Institutional promotion of this quality requires improving the status of teaching in higher education [4].

As has been mentioned in Chapters 2 and 8, a focus on higher aims and goals is inherent in the idea of the university. While a chiefly utilitarian perspective is sometimes prevalent in the higher education sector [5], the conventional expectation associated with scholarship, teaching, and learning, namely, their transformative potential in 'cultivating responsive and responsible individuals' [6], will continue to play a significant role. As discussed in this book, electromagnetics has both abundant utilitarian applications and a rich theoretical and conceptual framework. Together, these dual aspects provide a rich field for exploring ways of facilitating the overall development of students.

The following quote is relevant as we end this chapter in reminding us of higher principles connected with teaching and learning [6]:

> Higher learning can offer individuals and societies a depth and breadth of vision absent from the inevitably myopic present. Human being needs meaning, understanding, and perspective as well as jobs. The question should not be whether we can afford to believe in such purposes in these times, but whether we can afford not to.

Drew Faust (2009)

To extend the impact of the ideas in this book, additional materials are archived in the IEEE Antennas and Propagation Resource Center at https://resourcecenter.ieeeaps.org/. The page https://www.ieeeaps.org/education/educational-resource provides information on how to contribute to the Resource Center. We hope to further a conversation among teachers and students of EM theory and applications. Please contact the authors for further information, comments, or discussion.

REFERENCES

1. K. Moore, C. Jones and R.S. Frazier, "Engineering education for generation Z," *American Journal of Engineering Education*, vol. 8, no. 2, pp, 111–123, Dec. 2017.
2. M.J. Povinelli and J.A. Robinson, "*Integrating design thinking into an experiential learning course for freshman engineering students,*" *2018 ASEE Annual Conference & Exposition*, Salt Lake City, USA, June 24-27, 2018. Available at: https://www.asee.org/public/conferences/106/papers/22462/view, accessed November 23, 2020
3. K.T. Selvan, "Lessons learned from the IEEE AP-S Madras Chapter on electromagnetics education in India," *IEEE Antennas and Propagation Magazine*, vol. 63, no. 1, pp. 97–102, February 2021.

4. W.L. Edwards, *"Seeking excellence in higher education teaching: Challenges and reflections," Keynote address, 9th Intenrational Conference on University Learning and Teaching,"* November 28–29, 2018, Shah Alam, Malaysia. Available online at: https://files.eric.ed.gov/fulltext/EJ1207750.pdf, accessed November 23, 2020.
5. G. Boulton and C. Lucas, *"What are universities for?,"* September 2008. Available at: https://www.leru.org/files/What-are-Universities-for-Full-paper.pdf, accessed November 23, 2020
6. F. Kuriakose, "Possibilities of a professor: An academic in the twenty first century," *Higher Education for the Future*, vol. 5, no. 2, pp. 162–177, July 2018

Index

Printed in the United States
by Baker & Taylor Publisher Services